Incremental Data Converters for Sensor Interfaces

Incremental Data Converters for Sensor Interfaces

Chia-Hung Chen
National Yang Ming Chiao Tung University
Hsinchu, Taiwan

Gabor C. Temes
Oregon State University
Corvallis, OR, USA

IEEE PRESS
WILEY

Published by John Wiley & Sons, Inc., Hoboken, New Jersey.
Published simultaneously in Canada.

Trademarks: Wiley and the Wiley logo are trademarks or registered trademarks of John Wiley & Sons, Inc. and/or its affiliates in the United States and other countries and may not be used without written permission. All other trademarks are the property of their respective owners. John Wiley & Sons, Inc. is not associated with any product or vendor mentioned in this book.

Limit of Liability/Disclaimer of Warranty: While the publisher and author have used their best efforts in preparing this book, they make no representations or warranties with respect to the accuracy or completeness of the contents of this book and specifically disclaim any implied warranties of merchantability or fitness for a particular purpose. No warranty may be created or extended by sales representatives or written sales materials. The advice and strategies contained herein may not be suitable for your situation. You should consult with a professional where appropriate. Further, readers should be aware that websites listed in this work may have changed or disappeared between when this work was written and when it is read. Neither the publisher nor authors shall be liable for any loss of profit or any other commercial damages, including but not limited to special, incidental, consequential, or other damages.

For general information on our other products and services or for technical support, please contact our Customer Care Department within the United States at (800) 762-2974, outside the United States at (317) 572-3993 or fax (317) 572-4002.

Wiley also publishes its books in a variety of electronic formats. Some content that appears in print may not be available in electronic formats. For more information about Wiley products, visit our web site at www.wiley.com.

Library of Congress Cataloging-in-Publication Data Applied for:

Hardback ISBN: 9781394178384

Cover Design: Wiley
Cover Image: © Xinzheng/Getty Images

Set in 9.5/12.5pt STIXTwoText by Straive, Chennai, India

Contents

About the Authors

Chia-Hung Chen received his B.S. degree in nuclear engineering from National Tsing Hua University, Hsinchu, Taiwan, in 1994, and his M.S. degree in electrical engineering from Columbia University, New York, in 2003. He received his Ph.D. from Oregon State University, Corvallis, OR, in 2013.

Dr. Chen worked at start-up companies in Taiwan from 2003 to 2009, designing audio codec circuits, DC–DC power converters, and phase locked loops (PLLs). In 2013–2014, he was with MediaTek, Woburn, MA. From 2016 to 2018, he was a technical manager at EgisTec, designing fingerprint sensors and ADCs. He is currently an assistant professor at the Department of Electrical and Computer Engineering, National Yang Ming Chiao Tung University, Hsinchu, Taiwan. He has been serving as a member of the technical program committee (TPC) for the IEEE Custom Integrated Circuits Conference (CICC) since 2021 and VLSI Symposium on Technology, Systems and Automation (VLSI-TSA) since 2022. His research interests are in the design of precision analog circuits and energy-efficient data converters.

Gabor C. Temes (SM'66–F'73–LF'98) received undergraduate degrees from the Technical University of Budapest and Eötvös University, Budapest, Hungary, in 1952 and 1955, respectively. He received PhD degree in electrical engineering from the University of Ottawa, ON, Canada, in 1961, and an honorary doctorate from the Technical University of Budapest, Budapest, Hungary, in 1991.

Dr. Temes was Editor of the IEEE Transactions on Circuit Theory and Vice President of the IEEE Circuits and Systems (CAS) Society. In 1968 and 1981, he was cowinner of the IEEE CAS Darlington Award, and in 1984 winner of the Centennial Medal of the IEEE. He received the Andrew Chi Prize Award of the IEEE Instrumentation and Measurement Society in 1985, the Education Award of the IEEE CAS Society in 1987, and the Technical Achievement Award of the IEEE CAS Society in 1989. He received the IEEE Graduate Teaching Award in

1998, and the IEEE Millennium Medal, as well as the IEEE CAS Golden Jubilee Medal in 2000. He was the 2006 recipient of the IEEE Gustav Robert Kirchhoff Award and the 2009 IEEE CAS Mac Valkenburg Award. He received the 2017 SIA-SRC University Researcher Award. He is a member of the US National Academy of Engineering and the US National Academy of Inventors.

Preface

Sensors are playing a rapidly increasing role in our lives. They are used in instrumentation and measurement applications, in medicine as biosensors, in automotive controls, on the Internet of Things, in surveillance systems, in imagers, and in many other applications. The sensor outputs are physical analog signals and therefore require accurate analog-to-digital converters (ADCs) to enable their digital signal processing. Some sensor systems, such as imagers or neural recorders, contain dozens or even hundreds of sensors, each requiring data conversion. It is then desirable to share a single ADC among many of these sensors.

Historically, the accurate conversion of slow signals such as those measured by a digital voltmeter was performed by single- or dual-slope Nyquist-rate ADCs. These ADCs converted analog samples one by one into their digital equivalents. However, for even moderately fast signals, the long conversion time made these converters impractical. A noise filtering ("noise shaping") ADC, such as the delta-sigma ADC, provides fast and accurate conversion for streaming signals, but it cannot be used for measurement applications and cannot be multiplexed among several sensor channels. $\Delta\Sigma$ ADCs also suffer from several practical issues, such as the long latency between analog input and digital output, which makes them unsuitable for robotic and other feedback system applications. They also need complicated digital post filtering and may suffer from idle tones, instability, and other imperfections.

It was shown by R. J. van de Plassche in 1978 [1] that a first-order $\Delta\Sigma$ ADC can be converted into a highly accurate Nyquist-rate one by periodically resetting its analog and digital memory elements. He used bipolar current-mode circuitry in the realization of his new scheme with two chips. There was no immediate follow-up to his paper. Nine years later, an MOS-integrated version of the modified first-order $\Delta\Sigma$ ADC was implemented with switched-capacitor stages [2], and a year later, a second-order cascade realization was published [3]. Since the digital word was obtained by collecting increments of the input signals, the early authors named the device an incremental analog-to-digital converter (IADC). After these papers appeared, the many potential applications of this new type of converter became widely recognized. Soon, hundreds of papers were published which described a variety of innovative architectures and circuits. These extended the speed and accuracy of these IADCs. This resulted in important new applications and thus in the fabrication of many millions more IADCs both in embedded and stand-alone forms.

The theory and design techniques of IADCs have now been discussed in many research and tutorial papers and even book chapters [4, 5]. In spite of this exposure, the authors feel that the subject material is now extensive and mature enough to be exposed in a book of its own, which collects this information and can explain the theory and the many practical aspects of IADC design in sufficient detail.

The book is primarily aimed at industrial engineers, especially those involved in the design of analog- and mixed-mode (analog/digital) integrated circuits. However, the authors also used its material to teach short courses for industry, tutorial lectures, and graduate courses.

We thank the reviewers for their valuable suggestions and the contributions of Dr. Yi Zhang of Analog Devices Inc. to this work.

Hsin Chu, Taiwan, 01 September 2023   *Chia-Hung Chen*
Corvallis, Oregon, 01 September 2023   *Gabor C. Temes*

References

1 R. J. van de Plassche, "A sigma-delta modulator as an A/D converter," *IEEE Transactions on Circuits and Systems*, vol. 25, no. 7, pp. 510–518, July 1978.

2 J. Robert et al., "A 16-bit low-voltage A/D converter," *IEEE Journal of Solid-State Circuits*, vol. 22, no. 2, pp. 157–163, April 1987.

3 J. Robert and Ph. Deval, "A second-order high-resolution incremental A/D converter with offset and charge injection compensation," *IEEE Journal of Solid-State Circuits*, vol. 23, no. 3, pp. 736–741, March 1988.

4 C.-H. Chen, Y. Zhang, T. He, and Temes, G. C., "Micro-power incremental analog-to-digital converters," in *Efficient Sensor Interfaces, Advanced Amplifiers and Low Power RF Systems*, Springer, 2015. https://doi.org/10.1007/978-3-319-21185-5_2.

5 S. Pavan, R. Schreier, and G. C Temes, "Incremental analog-to-digital converters," in *Understanding Delta-Sigma Data Converters*, Wiley, 2017.

Abstract & Keywords

Abstract

Integrated sensor systems-on-chip require high-accuracy, low-latency analog-to-digital converters (ADCs) which can interface with wide input-range signals and often be shared among multiple sensor channels. Incremental analog-to-digital converters (IADCs) are often the best choice for such role. They are obtained by adding a simultaneous reset to the analog modulators and digital filters of a $\Delta\Sigma$ ADC. They are *Nyquist-rate* ADCs which use noise shaping to convert a finite number of analog samples into a single digital word. They retain most advantages of the $\Delta\Sigma$ ADCs and are much easier to be multiplexed, with shorter latency and simpler digital filters. In this book, the operation and design theory of IADCs is discussed. The advantages and disadvantages of IADCs and $\Delta\Sigma$ ADCs are also explained. It is shown how $\Delta\Sigma$ modulator architectures, such as multistage noise shaping (MASH), can also be used in the design of IADCs. A hybrid approach which combines an IADC with a Nyquist-rate ADC is introduced to further leverage the advantages of reset operation. The analysis and design of extended counting and hardware recycling schemes are then discussed. Five design examples are discussed, including a single-loop IADC, a multistep IADC, an IADC with multislope extended counting, a hybrid continuous-time IADC with a SAR, and a multistage multistep IADC.

Keywords

delta sigma ($\Delta\Sigma$) modulator
low latency
incremental analog-to-digital converters
multistage
multistep

extended counting
decimation filter
sensor interface
measurement and instrumentation

1

Fundamentals of Analog-to-Digital Data Converters (ADCs)

Sensors are devices which convert physical phenomena (sound, light, temperature, and others) into another signal, usually an electrical one. There are hundreds of applications for sensors in measurements and instrumentation, biomedical and environmental applications, Internet of Things (IoT), image sensors, and many more. In most cases, the electric output of the sensor is transmitted to a computer through an *analog interface*. This interface (often called *analog front-end or* AFE) may be used to amplify the sensor's output signal and to filter out unwanted noise from it. In most cases, it is followed by an *analog-to-digital data converter* (ADC) used to convert the AFE output into a digital form suitable for digital signal processing by a follow-up computer. The detailed structure of the AFE depends on the properties of the sensor output signal and on the application of the sensor. Figure 1.1 illustrates the block diagram of a sensor and its AFE. In this chapter, the fundamental principles of ADC are discussed, and an introduction to some high-accuracy data converters will be given.

1.1 Performance Parameters for Analog-to-Digital Converters

The ADC is often the most complex and critical part of the signal chain. Its specifications, as for those of the AFE, may vary widely, depending on the sensor signal and on the application of the device. Figure 1.2 illustrates the operation of an ADC. The input is an analog signal V_{in}, while the digital output D_{out} is a sequence of numbers which is the digital representation of V_{in}. The input–output relation is

$$V_{ref} \cdot D_{out} = V_{in} + V_{q} \tag{1.1}$$

Here, V_{ref} is the *reference voltage* of the converter, and V_{q} is the *quantization error*. The quantization error cannot be avoided, since V_{in} may take on *any value*

Incremental Data Converters for Sensor Interfaces, First Edition. Chia-Hung Chen and Gabor C. Temes.
© 2024 The Institute of Electrical and Electronics Engineers, Inc. Published 2024 by John Wiley & Sons, Inc.

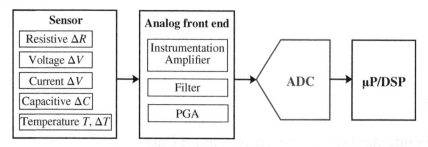

Figure 1.1 Analog front-end for sensor interface.

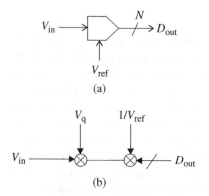

Figure 1.2 (a) The symbol of an ADC; (b) a simple ADC model.

within its range, while the digital signal is the sum of its bits (binary-weighted digits), and hence, it can only assume a *finite number of values*. The symbol of the ADC is shown in Figure 1.2, along with a simple model based on Eq. (1.1).

Figure 1.3 illustrates the normalized input–output characteristics of two M-level ADCs. $V_{ref} = 1\,V$ is assumed. Both ADCs are *bipolar*, i.e. able to convert both positive and negative inputs. Both have M steps and $M+1$ levels. The *resolution* of an M-step converter in bits is given by $N = \log_2 (M+1)$. The 45° line $k \cdot y$ shows the accurate output values which an infinite resolution ADC would provide for $V_{ref} = 1\,V$. The *least significant bit* value in the figure is $V_{LSB} = \Delta = 2$. The figure shows that in a range of the input range $-(M+1) < y < (M+1)$, the magnitude of quantization error $e = v - y$ satisfies $|e| < \Delta/2 = 1$. This is the *linear input range* of the ADC.

The difference between the two ADCs shown in Figure 1.3 lies in the location of the origin on the curves. For *the mid-rise quantizer*, it lies at a transition point; for *the mid-tread converter*, it lies in the middle of a flat portion (tread) of the curve. This difference may make the choice between the two options often obvious. The mid-tread ADC is less sensitive to noise, which is often an important advantage. However, if the ADC is used as a *quantizer* in a feedback loop (as is

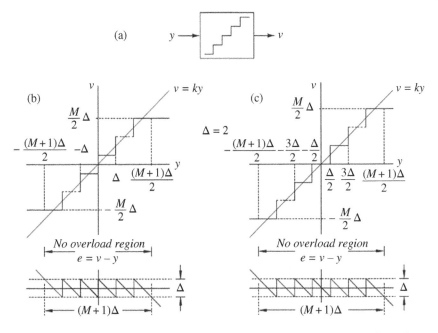

Figure 1.3 Normalized analog-to-digital converter transfer and error curves for a bipolar *M*-step ADC: (a) symbol; (b) curves for a mid-rise ADC; and (c) curves for a mid-tread ADC. The least significant bit value is $V_{LSB} = \Delta = 2$, and the slope is $k = 1/V_{ref} = 1$.

the case for a delta-sigma or incremental ADC), for very small input signals the mid-tread converter will not be able to change its output from zero, and an undesirable "dead zone" is created in the over-all transfer function.

Clearly, the conversion error $V_q = y - v$ is a *causal* variable, which can be found exactly from the ADC characteristic and the analog input in every clock period by analysis or simulation. However, to get a fast estimate of the expected performance, often we are treating the error as a *random white noise with a zero mean*. Its assumed mean square value can be derived by presuming that the probability of the error values outside the range $-V_{LSB}/2 < V_q < V_{LSB}/2$ is zero, and within that range it has a constant value. These approximations will be valid if the analog input of the quantizer varies sufficiently rapidly, so that the output code changes in almost every clock period. Under these conditions, the mean square value of V_q is given by

$$\sigma_q^2 = V_{LSB}^2/12 \tag{1.2}$$

The mean square value of V_{in} of a full-scale sine-wave signal is $\sigma_S = \dfrac{(2^N V_{LSB})^2}{8}$, and therefore, the signal-to-quantization-noise ratio (SQNR) is

$$SQNR = 10 \cdot \log\left(\sigma_s^2/\sigma_q^2\right) = 6.02\,N + 1.76\,(dB) \tag{1.3}$$

In addition to the SQNR, there are several parameters which can be used to characterize the performance of an ADC. These include its *zero error* and *gain error*. The zero error is the error of the first transition voltage in the input–output characteristic of the ADC. The gain error is the error of the difference between the first and last transition voltages.

Other ADC performance parameters are the *differential* and *integral nonlinearities* (DNL and INL). The DNL is the largest error in the analog step size which can generate a transition in the digital output. Its ideal value is V_{LSB}. The INL is the largest deviation of the characteristics from a straight line drawn from the lowest to the highest value of the characteristics. Notice that in finding the INL and DNL, we disregard the zero and gain errors.

A key performance parameter is the *signal-to-noise plus distortion ratio* (SNDR). It is the ratio of the signal power to the total noise-plus-distortion power. It can be found from

$$\text{SNDR} = 10 \cdot \log \left[\sigma_s^2 / \left(\sigma_q^2 + \sigma_n^2 + \sigma_d^2 \right) \right] \tag{1.4}$$

Here, σ_s^2, σ_n^2, and σ_d^2 are the mean square values of the signal, the noise, and the harmonic distortion, respectively. The total mean square error is thus a combination of the errors due to the quantization, the noise, and the nonlinear effects.

The *resolution* of the converter is the number of bits in its output. As shown in Eq. (1.3), it determines the SQNR of the ADC under ideal conditions. An artificially defined quantity, the *effective number of bits* (ENOB) is often used to characterize the performance of the nonideal ADC. The ENOB is the resolution (number of bits) of a *fictitious converter*, which has the same SNDR as the actual one *but is subject only to quantization error*. It can be obtained from Eq. (1.3) as ENOB = (SNDR − 1.76)/6.02.

Yet another important characteristic of the ADC is its *spurious free dynamic range* (SFDR). For a sine-wave input signal, the output spectrum of the ADC will contain a spectral line at the input frequency, and also other spurious lines caused by harmonic distortion, intermodulation, and other nonlinear effects. The difference between the signal and the largest spurious line, expressed in dB, is its SFDR.

1.2 Algorithms and Architectures for Analog-to-Digital Converters

There are many methods for performing ADC, each suited for a different application. ADCs can be divided into *memoryless* or *Nyquist-rate converters* and *memoried* or *oversampled ADCs*. The key feature of a memoryless converter is that it performs the conversion of each analog sample individually, independent of past

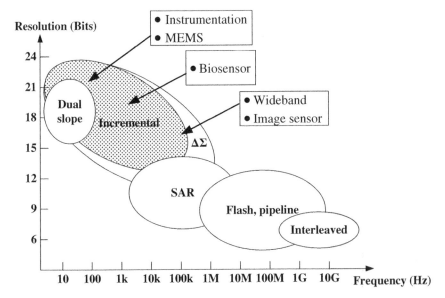

Figure 1.4 Operating regions of ADCs.

inputs. Thus, it is a *one-to-one conversion*. In a memoried ADC, the nth digital output $D(n)$ depends on the history of *all analog inputs* from the first (power-up) input $V_{in}(0)$ to the current one $V_{in}(n)$. Note that the "Nyquist-rate" converter cannot in fact sample the analog input at the true Nyquist rate f_N (which is defined as twice the signal bandwidth [BW]) without introducing aliasing. This is due to the imperfect antialiasing filter. Hence, the input sampling is often performed at two to four times the Nyquist rate. The *oversampled converters* may sample the input at a sampling rate f_s, which may be many times (hundreds of times) faster than f_N.

The many applications of ADCs led to many different implementations. They vary in their accuracy, speed, and power requirements. With practical limits on the complexity, power dissipation, and chip area of the circuit, a trade-off exists between speed and accuracy. Figure 1.4 illustrates the typical regions of performance for a few important ADC types.

There are many excellent textbooks which discuss most of these ADC realizations [1–3]. As Figure 1.4 shows, for high-accuracy ADC conversion of sensor signals, the available options include the *dual-slope, delta-sigma, and incremental* configurations. They will be briefly discussed next.

1.2.1 Dual-Slope (Integrating) ADCs

The block diagram of the dual-slope converter is shown in Figure 1.5. The input voltage V_{in} is integrated for N_1 clock periods, so the voltage V_x reaches the value

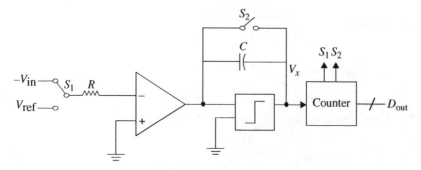

Figure 1.5 A dual-slope analog-to-digital converter.

$V_{in} \cdot N_1 \cdot T/(R \cdot C)$, where T is the clock period. Afterward, V_{ref} is integrated, causing V_x to drop by $V_{ref} \cdot T/(R \cdot C)$ in every clock period. This process is stopped after N_2 clock periods when V_x drops below 0, as detected by the comparator. The total drop is $V_{ref} \cdot N_2 \cdot T/(R \cdot C)$. Since the rise and drop are approximately equal, the relation $V_{in} \cdot N_1 \sim V_{ref} \cdot N_2$ results, giving $V_{in} \sim V_{ref} \cdot N_2/N_1$. Note that there is an error in this estimation of V_{in}, which can be as large as $V_{ref} \cdot T/(R \cdot C)$. To make this error less than the least-significant-bit (LSB) voltage for an N-bit conversion, $N_2 > 2^N$ is required. For high-accuracy ADC, this requires very long conversion time.

As the equations show, the conversion of V_{in} does not depend on the R and C values. However, it is affected by the linearity of the integration, and thus on the finite gain of the amplifier. It is also affected by its offset. Both effects can be mitigated by using circuit techniques, such as correlated double sampling.

A useful feature of the dual-slope ADC is that it can suppress some periodic noises, such as line noise ("hum") at its input. If the frequency of the noise is an integer multiple $1/(N_1 \cdot T)$, then it cancels during the input integration period.

The dual-slope ADC converts each input sample independently from the history of the ADC. It is thus a memory-less data converter.

1.2.2 Delta-Sigma A/D Converters

By allowing memory to play a role in the conversion process, the achievable accuracy within a fixed time period can be greatly enhanced. A popular architecture, which can achieve very high accuracy efficiently, is the *sampled-data delta-sigma* ADC (Figure 1.6a). The $\Delta\Sigma$ ADC consists of a feedback loop containing a sampled-data analog loop filter $H(z)$, a quantizer Q, and a D/A converter. The purpose of the loop is to force the spectrum of the digital output $D(z)$ to match that of the sampled analog input $V_{in}(z)$ in the signal band. To achieve this, the gain of $H(z)$ must be very large in this band. By treating the quantization error as an added

Figure 1.6 (a) The delta-sigma ADC; (b) Quantizer model.

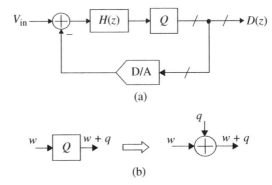

(a)

(b)

white noise $Q(z)$ (Figure 1.6b), and assuming that the digital-to-analog converter (DAC) is ideal with a reference voltage $V_{ref} = 1$, linear analysis can be used. It gives

$$D(z) = H \cdot V_{in}/(H+1) + Q(z)/(H+1) \tag{1.5}$$

Equation (1.5) confirms that at frequencies where $|H(z)| \gg 1$, $D(n) \sim V_{in}$, and $Q(s) \sim 0$. The choice of $H(z)$ is limited by the stability requirements of the feedback loop. Generally, outside of the signal band $|H|$ decreases, and hence, the quantization noise increases in the output.

In the system shown in Figure 1.6a, the quantizer and the DAC are necessarily sampled-data stages, while the input branch and the loop filter may be either sampled-data or continuous-time circuits. In the first case, linear analysis needs the use of the z-transform. Then Eq. (1.5) becomes

$$D(z) = H(z) \cdot V_{in}(z)/(H(z)+1) + Q(z)/(H(z)+1) \tag{1.6}$$

The quantization error $Q(z)$ depends on the input w of the quantizer and thus on the input signal $V_{in}(z)$. However, as discussed earlier, an estimate of the expected performance may be obtained by assuming that $Q(z)$ can be represented by an independent random signal, with its mean square value given by Eq. (1.2). We also assume that the mean value of quantization "noise" is zero, and its spectrum is white. Then we may define the *signal transfer function* (STF)

$$STF = D(z)/V_{in}(z)|_{Q=0} = H(z)/(H(z)+1) \tag{1.7}$$

and the *noise transfer function* (NTF)

$$NTF = D(z)/Q(z)|_{Vin=0} = 1/(H(z)+1) \tag{1.8}$$

In the following discussions, we shall assume that the input signal is in the baseband starting at DC, and its BW (Nyquist frequency f_B) is much lower than the clock rate. Then, by choosing $H(z)$ as a low-pass filter function with $|H(z)| \gg 1$ in the signal band, the desirable conditions $|STF| \sim 1$ and $|NTF| \ll 1$

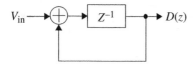

Figure 1.7 Sampled-date integrator (accumulator).

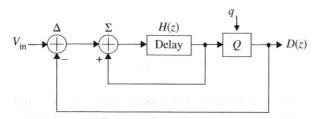

Figure 1.8 A first-order $\Delta\Sigma$ ADC (MOD1).

result from Eqs. (1.7) and (1.8). The simplest choice for a sampled-data filter is the accumulator shown in Figure 1.7. Its transfer function is

$$H(z) = z^{-1}/(1 - z^{-1}) \tag{1.9}$$

The resulting circuit of the $\Delta\Sigma$ ADC is shown in Figure 1.8. (This circuit explains the adjective $\Delta\Sigma$ given to this ADC, since it contains a differencing stage Δ followed by an accumulation one (Σ).)

Substituting the transfer function of Eq. (1.9) into the functions defined in Eqs. (1.7) and (1.8) gives

$$STF = z^{-1} \tag{1.10}$$

and

$$NTF = 1 - z^{-1} \tag{1.11}$$

Thus, the STF is simply a delay by a clock period T. The squared magnitude of the frequency response of the NTF is obtained by substituting $z = e^{j\omega T}$ and finding $|NTF|^2$. This results in

$$|NTF|^2 = [2 \cdot \sin(\pi f T)]^2 \tag{1.12}$$

Figure 1.9 shows the response of $|NTF|^2$ as a function of f.

To estimate the *in-band noise power* in the output signal, we need to integrate the power spectrum of the quantization noise at the output over the signal band 0 to f_B. Assuming that the noise is white, its power spectral density is $V_q^2 \cdot 2T = V_{LSB}^2 \cdot T/6$. The in-band noise power is then

$$N_q^2 = \int_0^{f_B} \frac{V_{LSB}^2 T}{6} 4 \sin^2(\pi f T) df \tag{1.13}$$

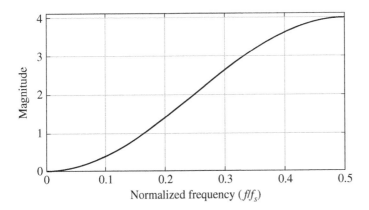

Figure 1.9 The frequency response of $|NTF|^2$.

Usually, the sampling frequency $f_s = 1/T$ is much higher than the Nyquist frequency f_B. Thus, the *oversampling ratio*, OSR $= f_s/2f_B$, is much larger than 1. Therefore, in the baseband the inequalities $f \cdot T < f_B/f_s \ll 1$ hold. This allows the use of the approximation $\sin^2(\pi f T) \sim (\pi f T)^2$ in Eq. (1.13), resulting in the relation given in Eq. (1.14).

$$N_q^2 \cong \frac{2}{3} V_{LSB}^2 T \int_0^{f_B} \pi^2 f^2 T^2 df = \frac{\pi^2}{36} \frac{V_{LSB}^2}{OSR^3} \tag{1.14}$$

Equation (1.14) shows that the in-band noise may be decreased by reducing V_{LSB} and increasing the OSR. Doubling the OSR reduces the in-band noise by 9 dB, and increases the SQNR by 1.5 bit. But the efficiency of this noise shaping is low. For a single-bit quantizer, it requires an OSR over 1000 to obtain an SQNR of 96 dB. To improve the effectiveness of noise shaping, higher-order loop filters may be implemented. Figure 1.10 shows a $\Delta\Sigma$ ADC MOD2 with a second-order loop filter. Linear analysis gives

$$STF = z^{-2} \tag{1.15}$$

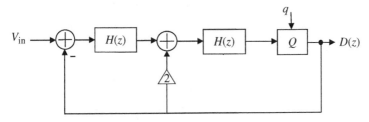

Figure 1.10 A second-order $\Delta\Sigma$ ADC.

and

$$\text{NTF} = (1 - z^{-1})^2 \tag{1.16}$$

The STF is now a delay by *two* clock periods $2T$. The squared magnitude of the frequency response of the NTF is obtained by substituting $z = e^{j\omega T}$. The result is

$$|\text{NTF}|^2 = [2 \cdot \sin(\pi f T)]^4 \tag{1.17}$$

The in-band quantization noise is now reduced by 15 dB, and the ENOB is increased by 2.5 bits if the sampling rate is doubled. An SQNR of 96 dB now requires only OSR = 128 for a single-bit quantizer. This is often acceptable.

The derivations provided for the first- and second-order delta-sigma ADCs may be generalized for the case of an Lth-order ADC with an N-bit quantizer, and a given OSR. The resulting formula (valid for OSR $\gg 1$) is given in Eq. (1.18):

$$\text{SQNR}_{\text{max}} \approx 6.02N + 1.76 + (2L + 1)10\log_{10}\text{OSR} - 10\log_{10}\frac{\pi^{2L}}{2L + 1}[\text{dB}] \tag{1.18}$$

The first term is the SQNR of the quantizer Q functioning alone, and the other two terms show the added contribution of the noise-shaping loop.

By increasing the order of the loop filter or the resolution of the quantizer of the $\Delta\Sigma$ ADC, and/or by increasing its OSR, the in-band noise may be reduced and the SQNR increased. However, there are inherent nonideal effects which will still limit the performance. The most important ones are listed below:

1. For $N > 1$, both differential and integral nonlinearities will occur in the DAC. This is due mostly to the mismatch errors of the DAC elements, and it causes harmonic distortion of the input signal. To mitigate this requires adding fairly complicated analog and digital circuitry, as shown in Chapter 2.
2. For $L > 2$, the stability of the feedback loop is not guaranteed and needs attention.
3. The power dissipation of the ADC increases approximately linearly with the OSR. The design of the active elements may also become difficult for high f_s.
4. The unavoidable correlation between the input of ADC and the quantization error neglected thus far may introduce large errors. These include the generation of *idle tones* and *dead zones*, to be described next.

There are various techniques available for mitigating the impact of these effects. Specifically, the linearity of the N-bit DAC may be improved by *dynamic element matching*. This is most easily applied to *unary DACs*, which are constructed by $2^N - 1$ equal-valued elements. For an input code of value D to the DAC then D unit elements are activated. By choosing these differently each time the same code appears, and averaging the results for many occurrences, the nonlinearity error may be converted into a white noise, or preferably into an appropriately filtered noise [4–6].

Figure 1.11   Dithering.

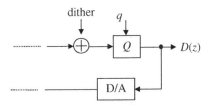

An alternative technique for mitigating DAC mismatches is *digital correction*. This acquires and stores the DAC errors for all possible input codes, and corrects them as they occur [7].

The stability of higher-order $\Delta\Sigma$ ADCs may be assured by an appropriate choice of the parameters of their NTF, such as their peak out-of-band gain. The process is discussed in Ref. [8] and will be described in Chapter 2.

Idle tones are generated for dc inputs which are rationally related to the quantizer's reference voltage, so that $V_{in} = (n/m)V_{ref}$. Here, n and m are integers and for stability $n < m$. As an illustration, assume $n = 1$ and $m = 4$, and let $V_{ref} = 1$ V. The quantizer outputs will be 0 or 1. Since the ideal NTF $= 0$ at dc, in steady state the output signal must have an average value of ¼. This will be the case if the sequence of the output values will be 0,0,0,1, 0,0,0,1,.... This output sequence is thus periodic, with a period of $f_s/4$. For high OSR, the tone generated will be out of the signal range, and hence, will not affect the signal-to-noise ratio (SNR) of the ADC. However, assume next that $n = 3$ and $m = 1000$. The tone generated now will have a frequency of $3f_s/1000$, which is likely to be in the signal band, greatly reducing the SNR.

Idle tone generation may be prevented by introducing a *dither* signal in front of the quantizer (Figure 1.11). This may be a random noise or an out-of-band signal. It is used to reduce or eliminate the causal relation between V_{in} and the quantization error q. To be effective, its amplitude should be larger than V_{LSB}.

Dead zones are input signal intervals, where the digital output does not follow variations of V_{in}. This effect is caused by leakage in the integrators of the loop due to the finite dc gain of the amplifiers used. It occurs when the charge provided by the input is less than the leaking charge.

Fortunately, tone generation and dead zones are both less likely to occur in higher-order $\Delta\Sigma$ ADCs and/or in the incremental ADCs which are the main subjects of this work.

1.2.3   Successive Approximation A/D Converters

The block diagram of an ADC with reduced accuracy but lower power and higher speed is shown in Figure 1.12 [1]. It is sometimes used as the quantizer of a $\Delta\Sigma$ or incremental ADC. The conversion scheme is based on gradually reducing the range which contains the estimated value of the analog input signal V_{in}.

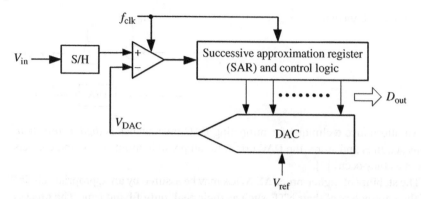

Figure 1.12 The successful approximation ADC. Source: Adapted from Carusone et al. [1].

For unipolar (positive only) conversion, assume that $0 < V_{in} < V_{ref}$. To find out whether V_{in} is in the upper or lower half of this range, it is compared with $V_{D/A} = V_{ref}/2$. The result of this comparison also provides the most significant bit b_1 of the output word. Next, it will be determined whether V_{in} is in the upper or lower half of reduced range. This can be found out by a second comparison with $(3/4)V_{ref}$ if the most significant bit (MSB) was 1, and with $(1/4)V_{ref}$ if it was 0. This reduces the range in which V_{in} must be by another factor of 2. Progressing this way for N clock periods, the N bits of the output word are obtained with an accuracy of $V_{LSB}/2$. This converter is often called *SAR ADC*, after the *successive approximation register* (SAR) used in its digital logic.

An efficient implementation of the successive approximation ADC is illustrated in Figure 1.13 for $N = 5$ bits [1]. It uses a binary-weighted capacitor array to carry out the conversion in three steps. In the *sampling step*, $V_{in} - V_{os}$ is stored in all capacitors, where V_{os} is the input offset of the opamp used here as a comparator. In the *holding* step, all capacitors hold their charge and voltage, but their common-mode voltage is shifted by $-V_{in}$. In the *conversion* phase, positive voltage steps are applied at the input of the opamp in order to compare V_{in} with fractions of V_{ref}. To find the MSB of the digital output, V_{in} is compared with $V_{ref}/2$. This is achieved by switching the bottom plate of the largest capacitor C from ground to V_{ref}. This generates a positive change by $V_{ref}/2$ in V_x. The sign of V_x determines the MSB b_1. Next, the modified output V_x is compared with $V_{ref}/4$. This is achieved by switching the bottom plate of second largest capacitor $C/2$ from ground to V_{ref}. This gives the next bit b_2. For 5-bit conversion, the process terminates after V_{in} is compared with $V_{ref}/16$, yielding the LSB.

If the comparator is realized using dynamic biasing, then all power dissipated is due to the charging and discharging of the capacitances in the capacitive DAC and the comparator. This allows the realization of the ADC with very little bias

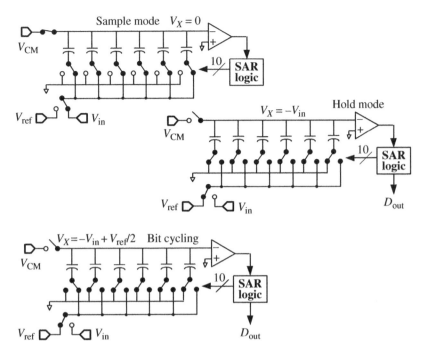

Figure 1.13 5-bit SAR ADC implementation. Source: Adapted from Carusone et al. [1].

power and makes the SAR ADC a very popular realization for micro-power A/D converters. By introducing oversampling, as well as additional capacitors and clock phases, noise-shaping SAR converters of improved accuracy can also be implemented [9, 10].

1.2.4 Flash A/D Converter

The fastest A/D converter is the appropriately named *flash ADC*. In this converter, shown in Figure 1.14 for $N = 3$ bits, the input voltage is simultaneously compared with all possible 2^N reference voltages V_{ri}. Assume that V_{in} is between V_{r3} and V_{r4}. Then the outputs of the bottom three comparators are negative, while those of the top five are positive. This information gives a digital output in the "thermometer" code. It can also readily be converted into a binary-coded digital equivalent of V_{in}.

The flash converter can be used as a stand-alone ADC. It is also often used as the quantizer of a $\Delta\Sigma$ or incremental ADC.

The complete conversion may be finished in 1 or 2 clock periods by a flash ADC, allowing low-resolution operation at very high (Gigasamples/second) conversion

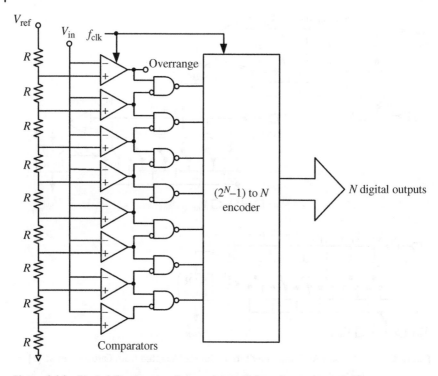

Figure 1.14 Flash A/D converter. Source: Adapted from Carusone et al. [1].

rates. The number of resistors and comparators, however, needs to be very high, around 2^N for N-bit resolution. This limits the practically achieved resolution to about 7 bits. The power dissipation is also high since the resistor string needs to recharge the input voltages of the comparators at a very fast rate. The fast charging operation of the flash ADC may also generate a large amount of noise on the chip, which may adversely affect the operation of analog circuits sharing chip area with the flash ADC.

1.2.5 Incremental A/D Converter

For very high conversion accuracy of low-frequency signals, $\Delta\Sigma$ ADCs are the most popular choice. However, there are applications where some of their shortcomings become critical. A few are listed below:

1. In systems which need many ADCs operating simultaneously, it may be wasteful to implement all of them separately. It then becomes economical to multiplex a single ADC between the channels. However, to multiplex a $\Delta\Sigma$ ADC all energy storage elements (capacitors in the analog loop, registers

in the digital filter) need to be replicated, and connected and disconnected when changing channels. A memoryless ADC by contrast may be multiplexed directly by simply moving it from channel to channel.

2. The digital filter needed in a $\Delta\Sigma$ ADC is usually highly complex since it is required to suppress a large amount of out-of-band quantization noise power. This requires extensive digital circuitry. Also, the high selectivity of the filter introduces a large delay (latency) between its input and output. It can be shown in Ref. [8] that for an Lth-order filter, this delay is at least $(L+1)T$, where T is the clock period after decimation. In ADCs used in feedback control systems, such large delay may destabilize the loop. For the modified version of the $\Delta\Sigma$ ADC described next, the digital filter may be designed in the time domain. This allows a simplified structure, and much reduced latency which is only one clock period T.

3. For lower-order $\Delta\Sigma$ ADCs the idle tone generation and dead zones may deteriorate the performance and may require additional circuitry to mitigate. For the modified ADC, the occurrence of idle tones and dead zones is reduced.

To convert the $\Delta\Sigma$ ADC into a memoryless ("Nyquist-rate"), the required modification is simple. The input is sampled, held for a given number of clock periods, and converted by the $\Delta\Sigma$ ADC and a digital filter. The digital result is a sample of the output. After this, all memory elements (capacitors in the analog circuits and registers in the digital ones) are discharged, a new input sample is acquired, and a new digital output sample is produced.

The modified $\Delta\Sigma$ ADC is often called an *incremental ADC* (IADC). Its properties and basic design principles are discussed in Chapters 2 and 3.

References

1 T. C. Carusone, D. A. Johns, and K. W. Martin, *Analog Integrated Circuit Design,* John Wiley and Sons, New York, 2012.

2 R. van de Plassche, *Integrated Analog-to-Digital and Digital-to-Analog Converters,* Kluwer Academic Publishers, Dordrecht, the Netherlands, 1994.

3 F. Maloberti, *Data Converters*, Springer, Dordrecht, the Netherlands, 2007.

4 R. T. Baird and T. S. Fiez, "Linearity enhancement of multibit A/D and D/A converters using data weighted averaging," *IEEE Transactions on Circuits and Systems-II*, vol. 42, pp. 753–762, Dec. 1995.

5 B. H. Leung and S. Sutarja, "Multibit σ-δ A/D converter incorporating a novel class of dynamic element matching techniques," *IEEE Transactions on Circuits and Systems-II*, vol. 39, pp. 35–51, Jan. 1992.

6 O. J. A. P. Nys and R. K. Henderson, "An analysis of dynamic element matching techniques in sigma-delta modulation," *Circuits and Systems*

Connecting the World. *ISCAS 96*, Atlanta, GA, USA, vol. 1, pp. 231–234, May 1996.

7 M. Sarhang-Nejad and G. C. Temes, "A high-resolution multibit $\Delta\Sigma$ ADC with digital correction and relaxed amplifier requirements," *IEEE Journal of Solid-State Circuits*, vol. 28, pp. 648–660, Jun. 1993.

8 Pavan, S., Schreier, R. and Temes, G.C., *Understanding Delta-Sigma Data Converters*, second edition, IEEE Press/Wiley Interscience, 2017.

9 J. A. Fredenburg and M. P. Flynn, "A 90-MS/s 11-MHz-bandwidth 62-dB SNDR noise-shaping SAR ADC", *IEEE Journal of Solid-State Circuits*, vol. 47, no. 12, pp. 2898–2904, Dec. 2012.

10 Y.-S. Shu, L.-T. Kuo and T.-Y. Lo, "An oversampling SAR ADC with DAC mismatch error shaping achieving 105 dB SFDR and 101 dB SNDR over 1 kHz BW in 55 nm CMOS", *IEEE Journal of Solid-State Circuits*, vol. 51, no. 12, pp. 2928–2940, Dec. 2016.

2

Delta-Sigma ADCs

Delta-sigma and incremental converters both rely on quantization noise shaping to achieve high accuracy. Hence, they have similar structures, with their main difference in the timing. There are also small differences associated with the termination and restarting of activities in the incremental analog-to-digital converter (IADC). In this chapter, we shall discuss the structures and operating principles of $\Delta\Sigma$ analog-to-digital converters. This is followed in Chapter 3 by the design of IADCs and a comparison of the two related converters.

As discussed in Chapter 1, a large quantization error can be reduced in the signal band by filtering, administered by a negative feedback loop. Figure 2.1 shows the basic structure used to realize the noise filtering. The input-referred replica of the quantization error Q is Q/H, and the output noise is $Q/(H+1)$. Hence, choosing H such as it assumes large values in the signal band will reduce the in-band quantization noise in the output by a factor of $1/|H|$. Using amplifiers in the filter H, this reduction can be very effective.

The filter used in the negative feedback loop may be a sampled-data (SD) circuit, usually in a switched-capacitor (SC) implementation, or a continuous-time (CT) one, usually realized by active-RC stages. Figure 2.1 assumes a SD implementation. The ADC of Figure 2.1 is called a $\Delta\Sigma$ structure, since it starts with a subtraction (Δ) at the input, followed by an accumulation (Σ).

The SD and CT implementations of the $\Delta\Sigma$ loop have complementary advantages and limitations. Thanks to the excellent matching accuracy of on-chip capacitors, the SC loop filter can implement its nominal transfer function precisely. However, those of its switches which operate at signal voltages are vulnerable to nonlinearities, due to signal-dependent charge injection and nonzero on-resistance. Also, the time limited slewing and settling of the SC integrators requires more power than the operation of a CT integrator at the same frequency. The operation of the CT loop filter requires matching of time constants of the form R·C or C/G_m. The implementation of these time constants is likely to introduce large errors, of the order of 20–40%. Such large errors will significantly change the

Incremental Data Converters for Sensor Interfaces, First Edition. Chia-Hung Chen and Gabor C. Temes.
© 2024 The Institute of Electrical and Electronics Engineers, Inc. Published 2024 by John Wiley & Sons, Inc.

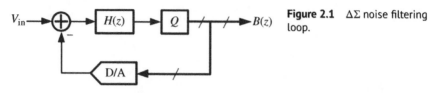

Figure 2.1 ΔΣ noise filtering loop.

noise transfer function (NTF) of the converter, and may even introduce instability. Hence, CT ΔΣ ADCs usually require tuning. On the other hand, they may use resistors, rather than switches at the signal-carrying nodes. These can operate linearly and they require less power for the same speed as SC ones. Also, they need simpler anti-aliasing prefiltering than their SC counterparts, or none at all. This is because in CT ΔΣ ADCs the sampling takes place after the CT loop filter, which therefore performs the anti-aliasing filtering.

Next, the properties of the SD ΔΣ ADCs will be discussed.

2.1 Sampled-Data ΔΣ ADCs

In this book, it will be assumed that the signal spectrum is located at the baseband, extending from dc to a limit frequency f_B, also called the *Nyquist frequency* f_N.[1] In this case, H will be realized by a high-pass filter. As mentioned in Chapter 1, the simplest SD ΔΣ ADC can be obtained by using a single accumulator (SC integrator) stage to implement $H(z)$. The resulting structure (Figure 2.2) is often called the MOD1 ΔΣ ADC [1]. Its signal transfer function (STF) is

$$H(z) = z^{-1}/(1 - z^{-1}) \tag{2.1}$$

The properties of the MOD1 ΔΣ ADC have been discussed in Chapter 1, and the second-order ΔΣ ADC (MOD2) shown in Figure 2.3 was also introduced. As described in Chapter 1, the NTF as well as the STF of the MOD2 are both the squares of those of MOD1. This allows a much higher signal-to-quantization noise ratio (SQNR) for the MOD2 than for the MOD1. The improvement can

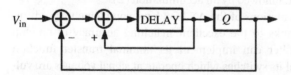

Figure 2.2 The first-order ΔΣ ADC (MOD1).

1 Note that strictly speaking the concept of the Nyquist frequency has only an approximate validity, since all signals involved must have a finite duration, which mathematically precludes a finite width for their frequency spectrum due to the uncertainty relation between the signal and its spectrum. However, for practical design, it is useful to arbitrarily assume a well-defined limit for the spectrum.

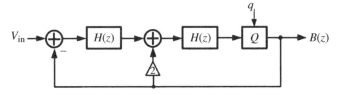

Figure 2.3 The second-order ΔΣ ADC (MOD2).

be predicted from Eq. (1.18) of Chapter 1. As expected, the improvement can be extended further by introducing MOD3, and so on. However, as will be discussed, the stability of the loop becomes compromised for orders $L > 2$, and the permissible input amplitude is also reduced.

In addition to raising the order L of the noise-shaping filter, the SQNR can also be enhanced by using increased resolution for the quantizer. Adding a bit to the resolution reduces the mean square value of the quantization noise by 6 dB. In fact, the potential SQNR improvement is even greater, since the reduced quantization error allows more aggressive loop design without introducing instability. The actual SQNR which can be expected for a given order L, quantizer resolution N and oversampling ratio (OSR) may be estimated from the computed curves [2, 3] shown in Figure 2.4.

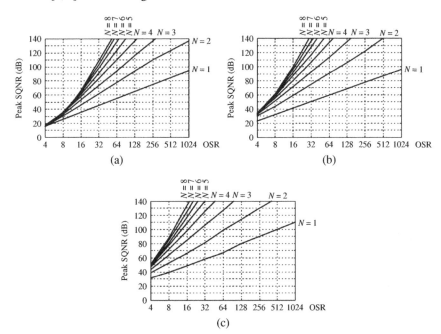

Figure 2.4 The estimated SQNR for ΔΣ ADCs: (a) with 1-bit quantizer; (b) with 2-bit quantizer; and (c) with 3-bit quantizer [2, 4].

These curves were obtained using the delta-sigma toolbox of MATLAB ([1], pp. 499–537). They assume the selection of optimized zeros and poles, to be discussed in Section 2.2 below.

2.2 Loop Filter Structures and Circuits for Sampled-Data $\Delta\Sigma$ ADCs

The architectures of MOD1 and MOD2 can be extended by cascading additional integrators and adding additional feedforward and/or feedback branches. The block diagram of MOD4 thus obtained is shown in Figure 2.5.

Following the notations of Refs. [1, 3], we may use *cascade of integrator* (CI) structures, with cascade of integrators and feedback (CIFB) and/or cascade of integrators with feedforward (CIFF). These configurations have different properties and different applications. Even for the same NTF, they may have different STFs.

A useful modification of the CIFF structure [4, 5] is illustrated in Figure 2.6, where the block providing $H(z)$ is stabilized by feedforward branches.

Linearized analysis of the structure gives

$$V = U + Q/(H + 1) \tag{2.2}$$

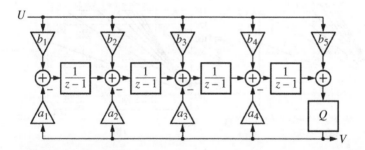

Figure 2.5 Block diagram of a MOD4 $\Delta\Sigma$ ADC.

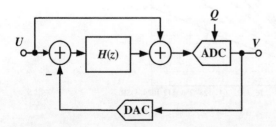

Figure 2.6 The low-distortion CIFF structure.

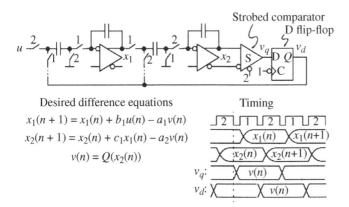

Figure 2.7 The simplified circuit diagram of a second-order ΔΣ ADC (MOD2).

Hence, the STF (under the linearized assumption and ideal component performance) *is equal to* 1. The input signal to the loop filter is

$$U - V = -Q/(H + 1) \qquad (2.3)$$

This does not contain the signal, only the filtered noise. This allows a significant relaxation of the linearity requirements of the loop filter, since a small nonlinearity only raises the noise floor somewhat, but does not introduce harmonics of the signal.

As mentioned earlier, in most cases SD ADCs are implemented using SC circuitry. As an example, Figure 2.7 shows the simplified circuit diagram of a second-order ΔΣ ADC, together with the difference equations valid for its states and the timing of its clock signals [1].

The MOD2 circuit is shown above in a single-ended realization. Infact, almost all actual implementations of ΔΣ ADCs are in a fully differential form, for suppressing even-order harmonics and for reducing external noise coupled into the circuit. The differential form of the first integrator of the MOD5 shown in Figure 2.5 is illustrated in Figure 2.8.

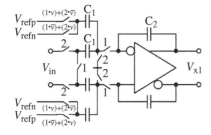

Figure 2.8 The differential circuit diagram of the first integrator of the MOD2 is shown in Figure 2.5.

2.3 Optimization of Zeros and Poles for Sampled-Data ΔΣ ADCs

Up to now, we have only discussed circuits with finite impulse response NTFs of the form $\text{NTF} = (1 - z^{-1})^L$. All L zeros of such NTF are at $z = 1$, i.e. at dc, and all its L poles are at $z = 0$. The resulting magnitude of the NTF gain response has a maximally flat behavior at zero frequency, similar to the gain response of a Butterworth filter. It is, therefore, expected that the in-band noise power can be reduced by introducing a ripple into the |NTF| versus frequency characteristic, making it more akin to the gain response of the more efficient Chebyshev filter. The optimum location of the L zeros can be obtained by minimizing the in-band power of the mean-squared noise [1, 3]. Figure 2.9 illustrates the $|\text{NTF}|^2$ frequency response for a MOD2 ΔΣ ADC with and without zero optimization. It indicates a large reduction of the noise power in the signal band $0 < \omega < 1$.

Table 2.1 shows the optimized zeros and the improvements obtained by using them [1]. As the table shows, the improvement rapidly increases with the order L of the ADC.

The zero improvement does not affect the FIR character of the NTF. A further modification, which may have an important effect on the stability of the feedback loop, is to change the NTF into an infinite impulse response (IIR) function. This is obtained by including a denominator polynomial in $\text{NTF}(z)$. This change shifts the poles away from the $z = 0$. The motivation for this was the observation that the stability and selectivity of the converter loop are sensitive functions of maximum value of $|\text{NTF}(z)|$ on the unit circle. This value is usually obtained at $z = -1$, corresponding to half of the clock frequency. By shifting the poles away from this point,

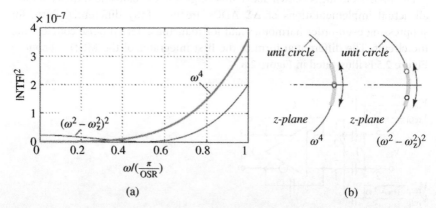

Figure 2.9 $|\text{NTF}|^2$ frequency responses for the MOD2 ΔΣ ADC with and without zero optimization: (a) frequency responses; (b) zero locations in the z plane [1].

Table 2.1 Optimum zero locations and SQNR improvements.

Order	Zero locations relative to band edge	SQNR improvement (dB)
1	0	0
2	$\pm 1/\sqrt{3} = \pm 0.577$	3.5
3	$0, \pm \sqrt{3/5} = \pm 0.775$	8
4	$\pm 0.340, \pm 0.861$	13
5	$0, \pm 0.539, \pm 0.906$	18
6	$\pm 0.23862, \pm 0.66121, \pm 0.93247$	23
7	$0, \pm 0.40585, \pm 0.74153, \pm 0.94911$	28
8	$0, \pm 0.18343, \pm 0.52553, \pm 0.79667, \pm 0.96029$	34

Source: Adapted from Pavan et al. [1].

the out-of-band gain (OBG) can be reduced. To preserve the shape of $|NTF(z)|$ in the signal band, the poles may be chosen as those of an Lth-order sampled-date Butterworth or Chebyshev filter. The MATLAB delta-sigma toolbox described in Refs. [1, 3] enables the user to find a suitable NTF.

2.4 Limitations on the Performance of Sampled-Data $\Delta\Sigma$ ADCs

The performance of the SD $\Delta\Sigma$ ADC is subject to linear as well as nonlinear errors. The causes of *linear errors* include mismatches between the capacitors in the SC loop filter, and the finite dc gain A_{dc} of the operational amplifiers. Capacitor mismatch errors can be minimized by proper layout, and the limited dc gain of the opamp does not have a large effect on the performance for second- or higher-order loops.[2]

An important potential source of *nonlinear distortion* is the signal-dependent errors introduced by the input sampling branches. These may be caused by the signal-dependent charge injection and nonzero on-resistance R_{on} of the floating switches. An efficient method of making these errors independent of the signal is to introduce *bootstrapping,* which makes the turn-on gate-to-source voltage V_{gs} always equal to a fixed bias voltage $V_{dd} - V_{ss}$. The concept is illustrated in

---

2 Note, however, that in low-order $\Delta\Sigma$ ADCs with low OSR the finite gain introduces *dead zones* at dc and at rational fractions of the reference voltage. In these, the output does not follow incremental changes of the input. This is a nonlinear error. But since the width of the dead zones is proportional to $(A_{dc}^{-L}) \cdot V_{ref}$, they have little effect for second- or higher-order ADCs.

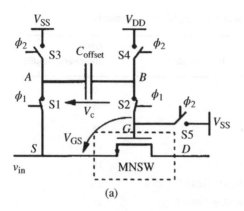

(a)

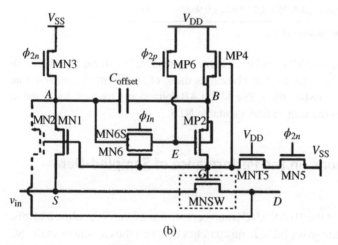

(b)

Figure 2.10 Bootstrapped switch (a) basic concept and (b) implementation [7].

Figure 2.10a. The added biasing capacitor C_{offset} is charged to $V_{dd} - V_{ss}$, and the gate of the input switch is grounded, during clock phase 2. During clock phase 1, C_{offset} delivers a constant gate-to-source voltage $V_{gs} = V_{dd} - V_{ss}$, resulting in signal-independent charge injection and R_{on}. To avoid the problem of turning on switch S1, and also any large internal voltages potentially causing breakdowns, the usual implemented circuitry tends to be quite involved [6, 7]. Figure 2.10b shows the circuit described in Ref. [7].

For ADCs with multi-bit quantizers, nonlinear effects are also caused by mismatch errors in the multi-bit feedback digital-to-analog converter (DAC). These cause signal-dependent noise entering the feedback loop and subtracted from the input signal. Hence, the overall linearity of the ADC is limited by the effective linearity of the DAC. By using dynamic element matching (DEM), the averaged effect of the DAC nonlinearity can be significantly mitigated. This process involves converting the output code D_{out} into a unary form and realizing the DAC from equal-valued least-significant-bit (LSB) unit elements. The unit elements are then used in rotating pattern, ensuring that their errors are averaged out over a long conversion period [1, 3, 8]. Alternatively, at the power-up, the DAC errors may be calibrated and stored in a look-up table (LUT). Then during conversion, for any output code, the error can be found in the LUT, and corrected [9].

For low-order ΔΣ ADCs with single-bit (binary) quantization, an important source of nonlinear distortion is the signal-dependent performance of the quantizer. Assume a dc input voltage $V_{ref}/2$ to a MOD1 with an initial integrator output signal of $V_{ref}/10$. Transforming Eq. (2.3) into the time domain, and using it to derive the output samples $v(n)$ gives the *periodic* sequence $v(n) = \{1, -1, 1, 1, |$ $1, -1, 1, 1, | ...\}$. The spectrum of $v(n)$ contains a dc term corresponding to $V_{ref}/2$ and also spurs at $f_s/4$ and its integer multiples. They are far out of band, and will be suppressed by the digital post filter. However, assume next that the dc input is at $V_{ref}/200$. The spurs will then occur at $f_s/400$ and its harmonics, which may well be in the signal band.

As already briefly discussed in Section 1.2.2, these spurs are called *idle tones* or *limit cycles*. They can be detected by entering a dc signal which is slowly varied in the permissible input range, and computing for each input value the in-band noise power. Figure 2.11 illustrates the results for a MOD1 for two OSR values.

Idle tones can be prevented by using dithering, illustrated earlier in Figure 1.11. As mentioned before, the dither may be a random noise or an out-of-band signal. In order to reduce or eliminate the causal relation between V_{in} and the quantization error q, the amplitude of the dither should be larger than V_{LSB}.

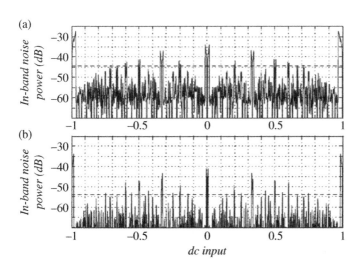

Figure 2.11 Idle tone patterns (a) for OSR = 32; (b) OSR = 64.

2.5 Multistage Sampled-Data ΔΣ ADCs

High-order $\Delta\Sigma$ ADCs are vulnerable to instability. It may be caused by the combination of the filtered quantization error and input signal overloading the quantizer. It is possible to derive a condition for absolute stability [10], but it is overly pessimistic. The theory of the $\Delta\Sigma$ ADC is very complicated, since it contains a nonlinear dynamic feedback structure. Hence, the stability is usually confirmed by extensive simulation. To avoid the stability issues raised by the high order of the feedback loop, and also to reduce the power required by the ADC, it may be constructed from a number of coupled stages. The resulting structure is called multistage noise shaping (MASH) ADC. The block diagram of the two-stage MASH is illustrated in Figure 2.12. The input of the added second stage is the quantization error e_1 of the first stage. It is converted into a digital form v_2, and digitally filtered by the H_2 stage. The two digital filters H_1 and H_2 are chosen so that in the overall output $v(n)$ the first-stage error e_1 is canceled, and is replaced by the filtered second-stage error e_2. Linear analysis in the z-domain results in the relation

$$V = H_1 \cdot V_1 - H_2 V_2 = H_1(\text{STF}_1 \cdot U + \text{NTF}_1 \cdot E_1) - H_2(\text{STF}_2 \cdot E_1 + \text{NTF}_2 \cdot E_2)$$

(2.4)

If the condition

$$H_1 \cdot \text{NTF}_1 = H_2 \cdot \text{STF}_2$$

(2.5)

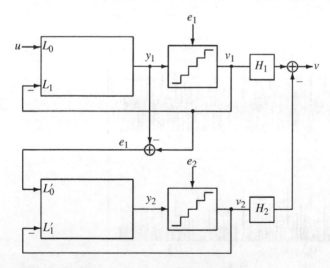

Figure 2.12 Block diagram of a MASH ΔΣ ADC.

holds, then E_1 is canceled in V. This can be achieved by choosing the digital transfer functions as

$$H_1 = \text{STF}_2 \tag{2.6}$$

$$H_2 = \text{NTF}_1 \tag{2.7}$$

Then the overall output becomes

$$V = \text{STF}_1 \cdot \text{STF}_2 \cdot U - \text{NTF}_1 \cdot \text{NTF}_2 \cdot E_2 \tag{2.8}$$

The STFs can usually be realized accurately for both stages, so the overall signal transmission is not affected by this cascading operation. However, the NTF is efficiently improved. For example, if the orders of both stages are chosen as $L_1 = L_2 = 2$, the order of the overall noise shaping is 4. Since both stages are second-order ones, the issues of stability and input signal overload are much relaxed.

However, conditions (2.6) and (2.7) must be exactly satisfied in order to prevent the "leakage" of e_1 into the overall output v. Since STF_2 can be accurately controlled, condition (2.6) can usually be met. However, the NTF_1 depends in a sensitive way on the opamp gains and element matches. In a single-stage $\Delta\Sigma$ ADC, the negative feedback at low frequencies in the loop desensitizes the output to these effects, but this is usually not enough to prevent noise leakage in the cancellation scheme used in MASH $\Delta\Sigma$ ADCs. Note that this cancellation requires the accurate matching of two time-varying signals H_2, a digital signal, and NTF_1, an analog one. It can be achieved by trying to make NTF_1 close to its ideal value or force H_2 to copy NTF_1. The latter approach was followed in Ref. [11].

As will be shown later, in the incremental ADC equivalents of the MASH ADC, only a single sample of the quantization error e_1 needs to be stored and canceled. This is a much easier task, which does not require extra care in the design of the first-stage loop.

2.6 Continuous-Time ΔΣ ADCs

As discussed earlier, SD $\Delta\Sigma$ ADCs allow the accurate realization of the targeted noise shaping, and thus allow high-resolution data conversion. However, they suffer from several problems. Since the input signal is sampled at the most sensitive nodes of the circuit, it must be kept free of aliased artifacts or nonlinear distortion. Thus, the circuit must be preceded by a CT anti-aliasing filter, and the sampling branches must be free of signal-dependent nonideal effects such as charge injection and settling errors. These requirements become even harder to meet with state-of-art technologies which restrict the supply voltages. Also, the input includes large spiking currents, which are difficult to generate for the preceding stage. Finally, the sampling input branch is subject to the "sampling penalty" increase of thermal noise, and hence, needs larger capacitors to overcome this.

Another disadvantage of the SD scheme is the requirement that the integrators used are forced to settle within a clock phase (half a clock period long or shorter) after receiving a fast input signal. This rapid settling requires a short slewing and settling time. This needs a large bias current, and therefore, large bias power for the opamps.

All of the listed shortcomings of the SD structure can be remedied, at some cost, by changing the structure of the loop filter into a CT one. In the CT implementation, the input branches as well as the feedforward and feedback branches may be realized with resistors. This allows simplicity and linear operation for these branches. The integrators can be realized by active-RC or G_m–C stages. The CT integrators operate with continuously varying input signals, and they do not need to change their outputs instantly. This allows them to operate with lower bias power (typically about 1/3 the power of the SC ones) for the same signal bandwidth and conversion accuracy. Figure 2.13 illustrates the block diagram of a CT MOD2 $\Delta\Sigma$ ADC [3].

The design of a CT $\Delta\Sigma$ ADC may be based on the established design methodologies of SD ones. Consider the block diagram shown in Figure 2.14.

Assume that the DAC is also realized with a CT circuit and that the quantizer is represented by its linearized model, with an error input $e(n)$. The only sampled component is then the switch at the ADC input. To calculate the loop gain of this mixed-mode structure, a model has to be found for the DAC. The two commonly

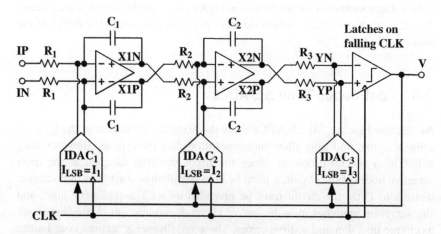

Figure 2.13 Block diagram of a CT MOD2 $\Delta\Sigma$ ADC.

Figure 2.14 CT ΔΣ ADC scheme.

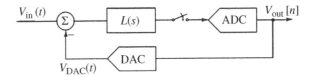

used CT DACs are the *non-return-to-zero* (NRZ) *circuit,* with an impulse response

$$p(t) = 1 \text{ for } 0 < t < T_s \tag{2.9}$$

and zero otherwise, and the *return-to-zero* (RZ) DAC, with impulse response

$$p(t) = 2 \text{ for } 0 < t < T_s/2 \tag{2.10}$$

$p(t)$ is zero outside of these ranges. Figure 2.15 illustrates the performance of these DACs.

The model of the loop is shown in Figure 2.16. The *s*-domain loop filter transfer function $L(s)$ should be found such that the CT circuit realizes the desired *z*-domain NTF(z) [1]. Note that by definition NTF(z) = $V_{out}(z)/E(z)|V_{in} = 0$. Here, $v_{out}(n)$ and $e(n)$ are both SD signals, so the NTF is well-defined even for CT ΔΣ ADC. For a prescribed NTF(z), the loop gain LG(z) may be found from the relation NTF(z) = $1/[1 + LG(z)]$.

The design of a low-order CT ΔΣ ADC from a specified NTF may start with assuming the block diagram of the circuit. It may have the same structure as the SD circuit providing the same NTF but with all blocks (integrators and coupling branches) replaced by their CT equivalents. Also, the scaling factors of all blocks

Figure 2.15 The impulse responses of the NRZ and RZ DACs.

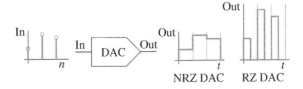

Figure 2.16 Closed- and open-loop models of a CT ΔΣ ADC.

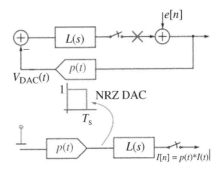

are different and must be determined by matching the actual LG(z) found from the circuit to the specified one.

To find the loop gain LG(z) from the block diagram, the loop is cut open after the switch, and the open loop is laid out as shown at the bottom of Figure 2.16. The loop gain LG(z) is then the transfer function of this branch.

As an example, we design a first-order CT ADC with NRZ DAC to provide the NTF of a MOD1, NTF(z) = $1 - z^{-1}$. It is expected that $L(s)$ will be a first-order function, realized by a CT integrator with a transfer function $I(s) = 1/s$, as shown in Figure 2.17. In the open-loop path, an analyis in the time domain of the model of Figure 2.17a gives the values of the signal samples in every clock period. Let the DAC input be $e(n) = \{1, 0, 0, \ldots\}$, the quantization error sequence. The resulting output of the NRZ DAC is $x_1(t) = u(t) - u(t - T_s)$, where $u(t)$ is the unit step function. The output of the integrator is $x_2(t) = t$ for $0 < t < 1$, and is equal to 1 for $t > 1$.

Figure 2.17 Closed-loop model of the CTMOD1 $\Delta\Sigma$ ADC.

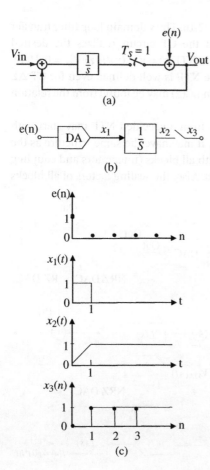

Figure 2.18 Design of a CTMOD2: (a) block diagram and (b) open-loop branch.

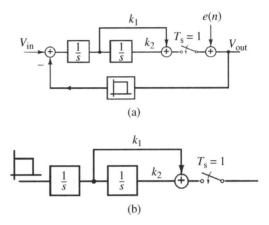

Finally, the sampled replica of $x_2(t)$ is the sequence $x_3(n) = [0, 1, 1, 1, \dots]$. The loop filter gain in the z-domain is the z-transform of $x_3(n)$:

$$LG(z) = z^{-1} + z^{-2} + z^{-3} + \cdots = z^{-1}/(1 - z^{-1}) \tag{2.11}$$

Hence, the NTF is

$$NTF(z) = 1/[1 + LG(z)] = 1 - z^{-1} \tag{2.12}$$

as required.

The same process [8] can be carried out for the design of a CTMOD2 ADC, for an NTF $= (1 - z^{-1})^2$. Figure 2.17 shows the assumed block diagram. Calculating the required LG(z) from the NTF, and equating it to the transfer function of the open-loop branch shown at the bottom of Figure 2.18 gives the coefficient values $k_1 = 1.5$ and $k_2 = 1$ [8].

2.7 Advantages and Limitations of Continuous-Time ΔΣ ADCs

An advantage of the CT ΔΣ ADC, not mentioned so far, is its inherent anti-aliasing performance. Assume that the input signal spectrum extends beyond the Nyquist limit $f_s/2$. Then at the sampling node nonlinear distortion is injected into signal, which cannot be corrected with any linear method. However, in the CT ΔΣ ADC sampling takes place at the same location as the quantization noise injection. Hence, the aliased artifacts are filtered by the NTF of the ADC and are effectively suppressed in the signal band. To obtain a more specific estimate of this anti-aliasing effect, consider the block diagram of Figure 2.19a. By moving the loop

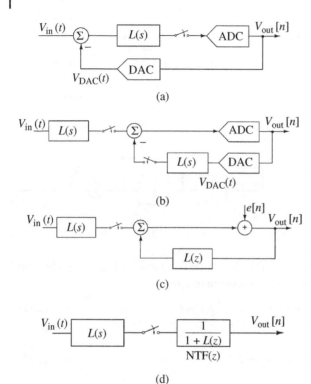

Figure 2.19 (a) Block diagram of the CT $\Delta\Sigma$ ADC; (b), (c), and (d) the transformed structures.

filter and sampling switch to the input leads of the adder, the equivalent structures of Figure 2.19b–d result.

In the last diagram, the second block has a transfer function $\mathrm{NTF}(z) = 1/[1 + L(z)]$. This is the target NTF of the CT $\Delta\Sigma$ ADC. It is typically of the form $\mathrm{NTF}(z) = (1 - z^{-1})^L$, where L is the order of the ADC. Then the STF can be written in the form

$$\mathrm{STF}(f) = L(j2\pi f).(1 - e^{-j2\pi f})^L \tag{2.13}$$

Figure 2.20 shows the magnitude response $|\mathrm{STF}(f)|$ for the CT MOD2 [1]. The inset shows the impulse response which corresponds to $\mathrm{STF}(f)$.

Figure 2.20 indicates the inherent anti-aliasing of the CT $\Delta\Sigma$ ADC. Input signals beyond the signal band are attenuated; signals with frequencies beyond f_s, which would be folded back into the signal band, are strongly suppressed before they are sampled.

Up to now, we have only discussed the advantages of CT $\Delta\Sigma$ ADCs with respect to the SD ones. However, there is no free lunch for ADC designers, and there are also severe problems which CT conversion introduces. The three key adverse

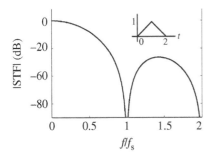

Figure 2.20 The signal gain response and impulse response of a CT MOD2 ADC.

effects are *the excess group delay, time constant variations, and imperfections in the DAC feedback signal.*

Excess loop delay (ELD) is unavoidable in any ΔΣ ADC loop, since it contains several delaying blocks: the ADC used as the quantizer, digital logic to perform DEM for multi-bit quantization, and the ADC. These blocks all introduce delays. In a SD ADC, the timing of all operations is well-defined, and the delays can be included in the design of the loop. In a CT ΔΣ ADC, the effect of the delays on the performance can also be predicted. As the theoretical analysis shows [1], an ELD $= t_d$ introduces an extra pole into the LG(z) function and causes the complex poles to approach the unit circle with increasing t_d, reducing the stability of the loop. To restore stability, additional zeros need to be introduced into LG(z), by adding feedforward or feedback branches to the loop. A simple structure applicable to the CTMOD2 ADC is shown in Figure 2.21. The coefficients k_0, k_1, and k_2 may be found by matching the impulse response of the realized loop filter, including the ELD, to the ideal one resulting in NTF(z) $= (1 - z^{-1})^2$. This results in three simultaneous equations for the k_i coefficients. Finding and applying these values restores the noise-shaping conditions of the CT ADC [1, 8].

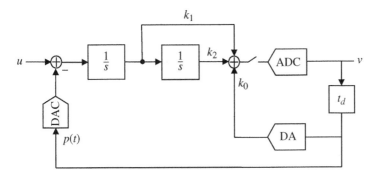

Figure 2.21 Excess loop delay correction in a CT MOD2.

In the SC implementation of a $\Delta\Sigma$ ADC, the time constants depend on the ratios of on-chip capacitor values and the clock frequency. With careful layout, these parameters can be implemented with very high accuracy, to within small fractions of 1%. In a CT $\Delta\Sigma$ ADC, by contrast, the time constants are given by $R \cdot C$ products or C/G_m ratios. The absolute errors of on-chip implementations of these values are around 15–30%, so these time constants have a large uncertainty. Since the frequency response of the loop filter is dominated by integrators, lowered time constants decrease the NTF in the signal band, and increase it out of band. The effect is similar to increasing the OBG, and it may lead to instability. Increased time constants decrease the noise-shaping effect, and allow more quantization noise in the signal band. Hence, both effects are harmful and need correction. This may involve tuning the time constants. There are several tuning methods available. A circuit which can perform continuous tuning of the resistors in the loop filter is shown in Figure 2.22.

The tuned resistors conduct current into the virtual grounds only when clock phase Phi is high. Hence, the average current is $d \cdot V_{pos}/R$, where d is the duty cycle of the clock signal. The output of the stage is fed into an unclocked comparator, which supplies clock phase Phi and determines the duty cycle d. The integrator output will settle at a value which satisfies $R/d = T_{ref}/C_{sc}$. Thus, the time constant of the tuned resistor and C_{sc} becomes T_{ref}. By using the same tuned clock signal for all resistors, and scaling all capacitors to C_{sc}, the time constants may be tuned to achieve good accuracy.

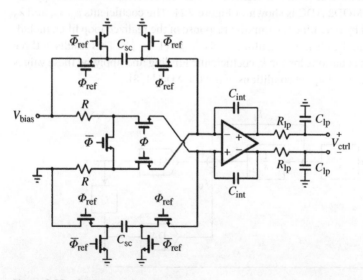

Figure 2.22 Resistor tuning circuit.

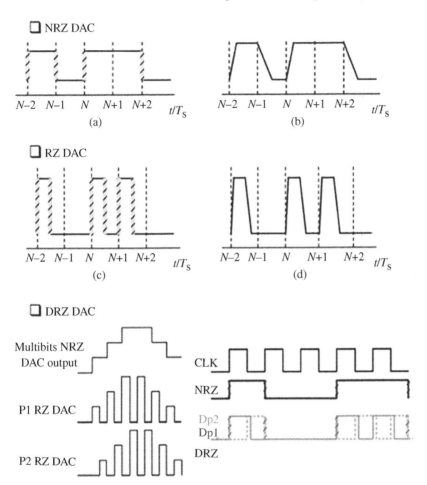

Figure 2.23 Jitter and ISI effects in NRZ, RZ and DRZ DAC signals. (a) Jitter in NRZ DAC, (b) ISI in NRZ DAC, (c) Jitter in RZ DAC, and (d) ISI in RZ DAC. Bottom: DRZ DAC signals.

The inaccuracy in the DAC signals is mostly due to *clock jitter* and *inter-symbol interference* (ISI). Clock jitter is due to random variations in the timing of the rising and falling edges of the clock signal (Figure 2.23). This modulates the length of the DAC signal, and hence, introduces a random noise. The longer the edges are, the larger this input noise is. Hence, its amplitude is about twice as large for the RZ DAC than for the NRZ one. ISI will be caused by the asymmetry of the rising and falling edges. Different slopes will cause an error at every transition. For RZ DACs, this error will be the same for every pulse, so it will cause only a dc offset. However, for an NRZ DAC, the occurrence of transitions is determined by the signal, and hence, ISI causes a *signal-dependent* noise, resulting in nonlinear distortion. A DAC scheme, which combines the advantages of both RZ and NRZ ones, is the *double-return-to-zero* (DRZ) DAC [12]. As illustrated at the bottom of Figure 2.23, this scheme uses two RZ DACs operated with a half-clock period delay. The resulting overall scheme is NRZ, so the jitter effect is reduced, but the individual waveforms are RZ, with IST causing only dc offset.

As discussed for SD $\Delta\Sigma$ ADCs earlier, for multi-bit DACs the mismatch errors of the DAC elements will also introduce nonlinear distortion, which may be mitigated by using DEM or calibration [1, 3, 5].

References

1 S. Pavan, R. Schreier, and G. C. Temes, *Understanding Delta-Sigma Data Converters*, second edition, IEEE Press/Wiley Interscience, 2017.

2 R. Schreier, "An empirical study of high-order single-bit delta-sigma modulators," *IEEE Transactions on Circuits and Systems II*, vol. 40, no. 8, pp. 461–466, Aug. 1993.

3 R. Schreier and G. C. Temes, *Understanding Delta-Sigma Data Converters*, IEEE Press/Wiley Interscience, 2005.

4 J. Silva, U. Moon, J. Steensgaard, and G. Temes, "Wideband low-distortion delta-sigma ADC topology," *Electronics Letters*, vol. 37, no. 12, pp. 737–738, Jun. 2001.

5 J. Steensgaard-Madsen, *High-Performance Data Converters*. Ph.D. dissertation, The Technical University of Denmark, 1999.

6 A. M. Abo and P. R. Gray, "A 1.5-V, 10-bit, 14.3-MS/s CMOS pipeline analog-to-digital converter," *IEEE Journal of Solid-State Circuits*, vol. 34, pp. 599–606, May 1999.

7 M. Dessouky and A. Kaiser, "Very low-voltage digital-audio $\Delta\Sigma$ modulator with 88-dB dynamic range using local switch bootstrapping," *IEEE Journal of Solid-State Circuits*, vol. 36, no. 3, pp. 349–355, Mar. 2001.

8 S. Pavan and N. Krishnapura, "Oversampling analog to digital converters," *21st International Conference on VLSI Design (VLSID 2008)*, Hyderabad, India, p. 7, 4 January 2008.

9 M. Sarhang-Nejad and G. C. Temes, "A high-resolution multi-bit $\Sigma\Delta$ ADC with digital correction and relaxed amplifier requirements," *IEEE Journal of Solid-State Circuits*, vol. SC-28, pp. 648–660, Jun. 1993.

10 J. G. Kenney and L. R. Carley, "Design of multibit noise-shaping data converters," *Analog Integrated Circuits and Signal Processing*, vol. 3, no. 3, pp. 259–272, May 1993.

11 P. Kiss et al., "Adaptive digital correction of analog errors in MASH ADCs. II. Correction using test-signal injection," *IEEE Transactions on Circuits and Systems, II*, vol. 47, no. 7, pp. 629–638, Jul. 2000.

12 R. Adams and K. Q. Nguyen, "A 113-dB SNR oversampling DAC with segmented noise-shaped scrambling," *IEEE Journal of Solid-State Circuits*, vol. 33, no. 12, pp. 1871–1878, Dec. 1998.

3

Single-Stage Incremental Analog-to-Digital Converters

Instrumentation and measurement applications usually require an analog-to-digital data converter (ADC) with high absolute accuracy and linearity at the analog interface between the sensor and the digital signal processor. Errors due to offsets and gain errors must also be very small. For conventional high-accuracy Nyquist-rate converters, such as dual-slope ADCs, large and accurate analog components, as well as high-gain precision amplifiers are necessary to achieve this. With delta-sigma ADCs, the signal-to-noise ratios (SNRs) can be very high for long stretches of signals, but high accuracy for individual samples is not guaranteed. Also, in some applications, such as imagers and electroencephalography (EEG), it is advantageous to multiplex a single ADC across multiple channels. There is no convenient way to accomplish this with delta-sigma ADCs. This chapter introduces the incremental analog-to-digital converter (IADC), which is particularly suitable for applications such as instrumentation and measurement. The rest of this book focuses primarily on this topic.

3.1 The First-Order IADC

IADCs are *Nyquist-rate* ADCs, which use oversampling and noise-shaping to accumulate a finite number of analog samples and convert them into a single digital word. Thus, the operation of IADC is best understood in the time domain [1–3]. Figure 3.1 shows the z-domain models of a delaying and a non-delaying sampled-data integrator with reset. When a reset pulse (RST) is used to discharge the integrator at the end of the oversampling window (when the time index is $i = M$), the output of the delay-free integrator can be found as

$$V[M] = \sum_{i=1}^{M} U[i] \tag{3.1}$$

Incremental Data Converters for Sensor Interfaces, First Edition. Chia-Hung Chen and Gabor C. Temes.
© 2024 The Institute of Electrical and Electronics Engineers, Inc. Published 2024 by John Wiley & Sons, Inc.

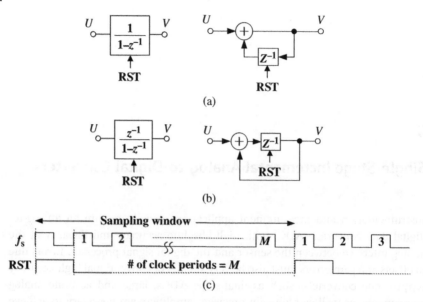

Figure 3.1 z-domain models of integrators with reset: (a) a delay-free integrator. (b) A delaying integrator. (c) Simplified timing in an oversampling window.

The integrator accumulates M samples of U and the average $\overline{U}$ can be defined as

$$\overline{U} = \frac{1}{M} \sum_{i=1}^{M} U[i] \tag{3.2}$$

A similar derivation gives for the output of the delaying integrator output and its average

$$V[M] = \sum_{i=1}^{M-1} U[i] \tag{3.3}$$

$$\overline{U} = \frac{1}{M-1} \sum_{i=1}^{M-1} U[i] \tag{3.4}$$

The integrators are shown in open-loop operation, which is also called "integrate-and-dump" [4]. They perform a finite impulse response (FIR) operation, which performs antialiasing filtering of the input signal [1–3].

Figure 3.2a depicts a first-order single-stage IADC (IADC1) comprised of an integrator and an internal quantizer in closed-loop operation. This analog circuit uses a low-distortion modulator with a direct input feedforward path. The simplified timing diagram, including the two-phase nonoverlapping clocks and reset pulse, is illustrated in Figure 3.2b. Using the oversampling frequency f_s and reset pulse RST at f_s/M, the operation of the IADC1 starts with a global

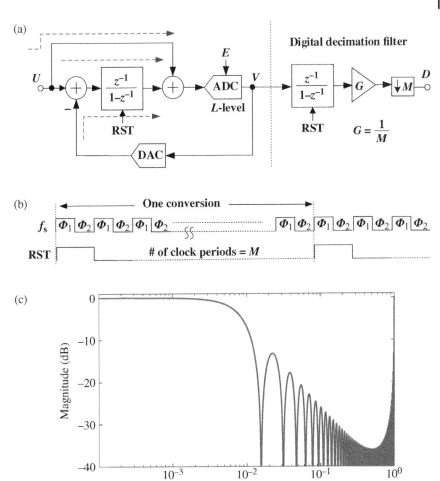

Figure 3.2 (a) The z-domain model of a single-stage first-order IADC with a low-distortion feedforward modulator. (b) The simplified timing diagram. (c) The signal transfer function (STF) of IADC1.

reset to clear the memories of both analog circuitry and digital circuitry. The *oversampling ratio* (OSR) M is defined as the number of oversampling clock periods within one conversion time period. At the end of the conversion time, when the time index is $i = M$, the signal at the output of the internal quantizer is

$$\sum_{i=1}^{M-1}(U[i] - V[i]) + U[M] + E[M] = V[M] \tag{3.5}$$

This can be rewritten as

$$\sum_{i=1}^{M} U[i] + E[M] = \sum_{i=1}^{M} V[i] \tag{3.6}$$

The average of the input signal $\overline{U}$ can be defined the same way as in Eq. (3.2). The closed-loop equation for one conversion can thus be written as

$$\overline{U} + \frac{1}{M} E[M] = \frac{1}{M} \sum_{i=1}^{M} V[i] \tag{3.7}$$

As Eq. (3.7) shows, the quantization error $E[i]$ which is introduced in every oversampled clock period is now averaged in the output. This is equivalent to a first-order noise-shaping. The digital decimation filter needed to reconstruct the digital equivalent D of the input from the oversampled digital bitstream V can be designed based on the right-hand-side of Eq. (3.7). The decimation filter can thus be as simple as an integrator-and-dump or simply as a counter. With a proper gain factor included, the reset pulse and registers will decimate and deliver the digital words D at the Nyquist rate f_s/M. After M clock periods, the next reset pulse reads the output word and clears all memories. The circuit clearly converts analog data *sample-by-sample*, and hence, functions as a *Nyquist-rate ADC*.

The IADC1 has an inherent advantage over most other ADC types in some sensor interface applications: the STF plotted in Figure 3.2c, shows a sin x/x type filtering whose frequency response has notches at integer multiples of the frequency f_s/M. This FIR filtering also occurs in slope-based ADCs, for example, in a dual-slope ADC. It is useful for suppressing periodic noise and interference (usually the line-frequency noise) at multiples of f_s/M. However, to achieve N-bit resolution, the oversampling clock frequency f_s must be 2^N times the Nyquist rate. Hence, IADC1 has a poor energy efficiency. Nevertheless, it finds applications in some sensor interfaces [5].

3.2 Higher-Order Single-Stage IADCs

3.2.1 Analysis and Design of a Second-Order IADC

The IADC1 can be extended to a second-order IADC (IADC2). Its design and operation are derived in this section. Figure 3.3 depicts the z-domain model of an IADC2 with a low-distortion feedforward modulator loop [6]. The timing diagram in Figure 3.2b can also be used for the IADC2. Again, M is the OSR, defined as the number of oversampling clock periods within one conversion period. After reset, the $\Delta\Sigma$ modulator loop quantizes the analog input voltage U, and the digital filter concurrently processes the output bitstream V.

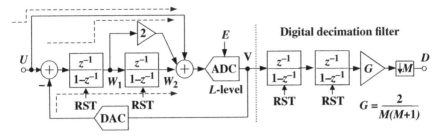

Figure 3.3 The z-domain model of a single-stage IADC2 with a low-distortion feedforward modulator.

In the feedback loop, both integrators accumulate the difference between the input signal U and the modulator's output V. At the end of the conversion period (time index $i = M$), the outputs at the first and second integrators are

$$W_1 = \sum_{i=1}^{M-1}(U[i] - V[i]) \tag{3.8}$$

and

$$W_2 = \sum_{K=1}^{M-1}\sum_{i=1}^{K-1}(U[i] - V[i]) \tag{3.9}$$

respectively.

The input and output of the internal ADC satisfy the equation

$$U[M] + 2 \cdot W_1[M] + W_2[M] + E[M] = V[M] \tag{3.10}$$

From Eqs. (3.8) and (3.9)

$$U[M] + 2\sum_{i=1}^{M-1}U[i] + \sum_{K=1}^{M-1}\sum_{i=1}^{K-1}U[i] + E[M] = V[M] + 2\sum_{i=1}^{M-1}V[i] + \sum_{K=1}^{M-1}\sum_{i=1}^{K-1}V[i] \tag{3.11}$$

$$\sum_{K=1}^{M}\sum_{i=1}^{K}U[i] + E[M] = \sum_{K=1}^{M}\sum_{i=1}^{K}V[i] \tag{3.12}$$

If the input $U[i]$ is constant during conversion, $\sum_{K=1}^{M}\sum_{i=1}^{K}U[i] = M(M+1)/2$. Hence, we may define the *average input voltage* $\overline{U}$ by the relation

$$\overline{U} \cong \frac{2}{M(M+1)}\sum_{j=1}^{M}\sum_{i=1}^{j}U[i] \tag{3.13}$$

Note that $\overline{U}$ represents the input accurately only if U does not vary significantly during the conversion. The least-significant-bit (LSB) quantization error E of the internal L-level quantizer is $V_{FS}/(L-1)$, where V_{FS} is the full-scale voltage. From Eq. (3.13), if the final error $E[M]$ is the LSB error, the input may be estimated from

$$\overline{U} + \frac{2}{M(M+1)}\frac{V_{FS}}{L-1} = \frac{2}{M(M+1)}\sum_{K=1}^{M}\sum_{i=1}^{K}V[i] \tag{3.14}$$

To reconstruct $\overline{U}$ from the output bit stream V, the digital decimation filter should perform the operation on the right-hand side of Eq. (3.14). For an IADC2, the decimation filter can thus be simply two counters in cascade (Figure 3.1). Alternatively, a multiply and accumulate (MAC) operation may be used. As Eq. (3.14) shows, the loop filter samples the input signal M times in one conversion, and performs a FIR filtering on the input signal.

The LSB quantization error E of the internal L-level quantizer is $V_{FS}/(L-1)$, where V_{FS} is the full-scale voltage. The equivalent LSB quantization error E_{IADC2} of the IADC2 is, therefore,

$$E_{IADC2} = \frac{2}{M(M+1)}\frac{V_{FS}}{L-1} \tag{3.15}$$

The effective number of bits (ENOBs) and the signal-to-quantization-noise-ratio (SQNR) at full-scale input amplitude are given by

$$ENOB_2 = \log_2\left(\frac{V_{FS}}{E_{IADC2}}\right) \tag{3.16}$$

Hence, the SQNR for a full-scale sine-wave input is

$$SQNR_2 = 10\,\log\left(\frac{V_{FS}^2/8}{E_{IADC2}^2/12}\right) = 2 \times 20\,\log(M) + 20\,\log(L-1) - 4.3 \tag{3.17}$$

3.2.2 The Design of Higher-Order IADCs

To design a higher-order loop filter, the design methodology of conventional $\Delta\Sigma$ ADCs may be used. It can be found in Chapter 2 of this book and Refs. [4, 6]. The design process includes finding the loop filter architecture, the loop filter order, the OSR, the zeros and poles, and quantizer resolution. A general $\Delta\Sigma$ ADC, shown in Figure 3.4, consists of an oversampled loop filter, a low-resolution quantizer, and a multi-rate decimation filter. N cascaded integrators form an Nth-order loop filter. The quantization noise spectrum is high-pass filtered by the noise transfer function (NTF), which is usually of the form

$$NTF(z) = \frac{V}{E} = \frac{1}{1+L_1(z)} = \frac{(z-1)^N}{a_L z^N + a_{L-1}z^{N-1} + \cdots a_0} \tag{3.18}$$

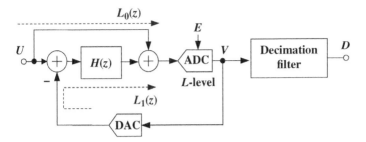

Figure 3.4 The conventional single-stage $\Delta\Sigma$ ADC.

Here, $L_1(z)$ is the loop transfer function. Schreier's toolbox [4] is a powerful help for finding the coefficients, optimizing the out-of-band gain (OBG), performing dynamic range scaling, and finding other practical design parameters.

Using Eqs. (3.2)–(3.14), it is straightforward to extend the analysis of an IADC2 to derive similar results for an Nth-order IADC (IADCN). The equivalent quantization error and the maximum SQNR of an IADCN are now

$$E_{\text{IADCN}} \approx \frac{N!}{M^N} \frac{V_{\text{FS}}}{L-1} \tag{3.19}$$

$$\text{SQNR}_N \approx N \cdot 20\log(M) + 20\log(L-1) - 20\log(N!) \tag{3.20}$$

The Nth-order loop filter scales down the internal quantization error by a factor $N!/M^N$. Figure 3.5 shows the calculated SQNR versus OSR for the modulator

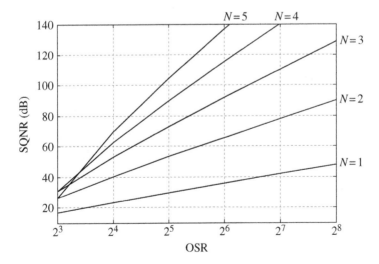

Figure 3.5 SQNR versus OSR for a single-bit ($L = 2$) modulator from first-order ($N = 1$) to fifth-order ($N = 5$) loops.

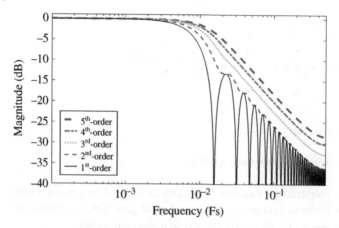

Figure 3.6 Signal transfer function of IADCs from first-order to fifth-order.

with a two-level ($L = 2$) internal quantizer from orders $N = 1-5$, assuming that the modulator is not overloaded at full-scale input voltage.

Figure 3.6 shows the simulated STFs of IADCs for first- to fifth-order loops. Note that the STF notches of an IADC1 no longer appear in higher-order IADCs. The STF still shows low-pass filtering, while the conventional feedforward $\Delta\Sigma$ modulators often have unity STF [6].

In addition to the low-pass filtering STF, the IADC has also much less low-frequency pattern noise, or idle tones, than its $\Delta\Sigma$ ADC counterparts [7]. However, there may be "dead bands" around the quantizer's thresholds (around zero for two-level quantizer). They can be eliminated by injecting dither [1].

3.2.3 IADC Circuit Techniques

Some circuit techniques incorporated in $\Delta\Sigma$ modulators can also be applied to IADCs. FIR feedback digital-to-analog converter (DAC) [8] is an effective way to reduce the integrator's voltage steps due to the DAC signal. The reduced voltage steps mitigate the effects of clock jitter, inter-symbol-intermodulation (ISI), and distortion in continuous-time integrators and also reduce the opamp's slewing in switched-capacitor integrators. Figure 3.7 shows an example of a feedforward IADC1 with a FIR DAC. In the modulator design, the filtered DAC output $F_{DAC}(z) = V(z) \cdot F(z)$ alters the loop filter transfer function $L_1(z)$ shown in Figure 3.4. This effect must be compensated by an additional path $C(z)$ to restore the NTF to its original form.

In the overall IADC operation, the added $F(z)$ signal must be added also in the digital filter ahead of the cascaded integrators to reconstruct the bitstream V

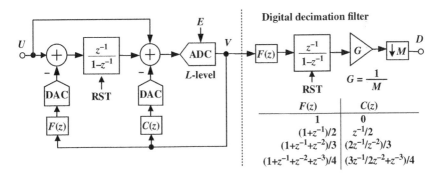

Figure 3.7 IADC1 with a FIR DAC.

for the correct D. Design examples are illustrated in Refs. [9, 10], where the FIR DAC is incorporated to reduce the voltage step, and thus the energy consumption, effectively.

3.2.4 Comparison of IADCs and $\Delta\Sigma$ ADCs

A comparison of the properties of an IADC and the corresponding $\Delta\Sigma$ ADC for the same conversion task shows the following:

Accuracy: comparison with the continuously operated $\Delta\Sigma$ ADC that uses the same circuit for the same input signal, shows lower accuracy for the IADC, due to the time lost in resetting its circuitry.

Stability and idle-tone generation: high-order $\Delta\Sigma$ ADCs may have stability issues; while low-order $\Delta\Sigma$ADCs may generate idle tones (limit cycles). The reduced active conversion time allows improved stability for the IADC, and also a decreased tendency to generate idle tones.

Decimation filter complexity and latency: as will be demonstrated shortly, the digital filter for the IADC can be much simpler than that for the $\Delta\Sigma$ADC. Hence, its *latency* (the time that elapses between the entry of an analog sample and the exit of the corresponding digital word) is significantly shorter than that of the $\Delta\Sigma$ ADC. Consequently, the IADC is a more advantageous choice for applications such as control systems or robotics where the ADC is part of a feedback loop.

Multiplexing: thanks to the IADC's ability to switch channels at reset times, it is naturally suited to multiplexing. By contrast, multiplexing a $\Delta\Sigma$ ADC requires the replication of all memoried elements (capacitors and registers), since it needs a different one for each channel. IADCs are therefore a natural choice for multichannel systems such as EEKGs and imagers.

3.3 Decimation Filter and the Overall Design of IADCs

3.3.1 Cascade-of-Integrators (CoI)

The decimation filter samples the digitized bitstream at oversampled-rate (f_s) and then converts it to digital words at Nyquist rate (f_s/OSR). The decimation filter for a conventional N^{th}-order $\Delta\Sigma$ ADC is shown in Figure 3.8a. It has several stages and usually consists of a sinc^{N+1} filter and an FIR half-band filter (HBF). The *Hogenauer* sinc^3 decimation filter [11] for a second-order modulator is shown in Figure 3.8b. The decimation filter in an IADC is much simpler. It may be a cascade of a few accumulators, or a single MAC stage. Thus, the energy efficiency is improved. This saves chip area, and the resulting sensor system-on-chip (SoC) will be cost effective.

In Figure 3.8, the conventional decimation filter introduces additional delay (latency) from bitstream input V to the final digital output D. If the oversampled bitstream is decimated down to Nyquist rate by a sinc^3 filter and no additional filtering is used, the latency becomes $3 \cdot T_{\mathrm{CONV}}$, where T_{CONV} is a required time for single Nyquist conversion period. In general, for a Nth-order $\Delta\Sigma$ modulator, the latency is at least $(N+1) \cdot T_{\mathrm{CONV}}$. Comparison of the IADC2 shown in Figure 3.3 with the conventional $\Delta\Sigma$ ADC in Figures 3.4 and 3.7 shows that the latency from the analog input to the decimated digital output of an IADC, which is only one Nyquist conversion period, is *at least $N + 1$ times shorter*. The short latency is essential when the ADC is incorporated into a closed-loop sensor or actuator system. There, any excess loop delay may bring instability, and the necessary compensation will complicate the system design.

3.3.2 Thermal Noise

The SNR of a $\Delta\Sigma$ ADC is usually limited by the sampled kT/C thermal noise. It is usual to assign the largest part of the noise budget for an IADC to thermal noise,

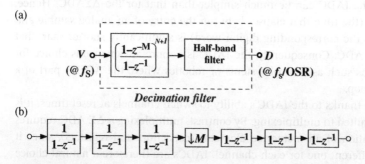

Figure 3.8 (a) The decimation filter of a conventional $\Delta\Sigma$ ADC. (b) A Hogenauer sinc^3 decimation filter for a second-order $\Delta\Sigma$ ADC.

since reducing it is more costly in power dissipation than reducing quantization noise. In a $\Delta\Sigma$ ADC, the input-referred noise power $P_{n,\,\text{in}}$ is $\overline{v_{kTC}^2}/M$. Here, $\overline{v_{kTC}^2}$ is the input noise power of each sampled kT/C noise source. However, when a higher-order $\Delta\Sigma$ ADC is operated as an IADC, every conversion has a *time-dependent* weight factor w_i associated with each input sample, and the total input-referred noise power is

$$P_{n,\text{in}} = \sum_{i=1}^{M} w_i^2 \overline{v_{kTC}^2} = \overline{v_{kTC}^2} \cdot \sum_{i=1}^{M} w_i^2 \tag{3.21}$$

For an IADC1, all the weight factors $w_1 = w_2 = \ldots = w_M = 1/M$ and w_i is an equal weight for all M samples. Then the IADC1 sampled thermal noise is simply $\overline{v_{kTC}^2}/M$.

In an IADC2 with two cascaded integrators used as a decimation filter, the weight factor w_i is

$$w_i = \frac{2}{M(M+1)} \cdot i \tag{3.22}$$

The IADC2's input referred noise power is now

$$\overline{v_{kTC}^2} \cdot \sum_{i=1}^{M} w_i^2 = \overline{v_{kTC}^2} \cdot \left(\frac{2}{M(M+1)}\right)^2 \cdot (1^2 + 2^2 + 3^2 + L + M^2) \approx \frac{4}{3M} \tag{3.23}$$

Thus, for $M \gg 1$, $\sum w_i^2$ reaches a maximum value of $1.33/M$ for a second-order IADC. The derivation can be repeated for higher-order IADCs, and $\sum w_i^2$ reaches $1.8/M$ for third-order IADC. For a fourth-order IADC, this *thermal noise penalty* can be 2. A more detailed analysis can be found in Refs. [1, 3, 12].

Since a higher-order IADC has increased thermal noise, the size of the input capacitor should be increased to maintain the same SNR. The amplifiers then need more power to drive their increased capacitive loads. Hence, higher-order IADCs consume more power than the equivalent $\Delta\Sigma$ ADCs. This noise penalty is part of the trade-off in using the IADC.

3.3.3 Optimized Digital Filter Design for a Single-Stage IADC

There is a more sophisticated method of designing the digital filter, that can minimize the thermal and quantization noises separately, or their weighted sum [12]. The IADC model used in the design assumes that the input thermal noise is white and that its mean square value is $\gamma k_B T/C_{\text{in}}$. Here, k_B is the Boltzmann constant, T *is* the temperature in degrees Kelvin, and γ represents a scale factor determined by the circuitry of the input branch [6]. Typically, γ equals 5. Then it can be shown that the mean square value of the output thermal noise can be calculated as follows:

$$P_t = \frac{\gamma k_B T}{C_{\text{in}}} \mathbf{h}^T \mathbf{S}^T \mathbf{S} \mathbf{h} \tag{3.24}$$

Here, $\boldsymbol{h}$ is a column vector of M elements, the kth element of which is $h[k]$, the kth sample of the impulse response of the decimation filter. Also, $\boldsymbol{S}$ is the triangular matrix

$$S = \begin{bmatrix} s[0] & 0 & 0 & \cdots & 0 \\ s[1] & s[0] & 0 & \cdots & 0 \\ \vdots & \vdots & \vdots & \ddots & \vdots \\ s[M-1] & s[M-2] & s[M-3] & \cdots & s[0] \end{bmatrix} \quad (3.25)$$

In this array, $s[k]$ represents the kth sample of the impulse response of the STF of the loop. For a low distortion loop, STF $= 1$, and $\boldsymbol{S}$ becomes the unit matrix. Then P_t is equal to $(\gamma k_{\text{B}} T / C_{\text{in}}) |\boldsymbol{h}|^2$. In order to minimize the thermal noise, $h[k]$ should be chosen so as to minimize P_t, while also ensuring that the dc gain of the digital filter is one. This gain condition can be translated into

$$\boldsymbol{e} \cdot \boldsymbol{h} = 1 \quad (3.26)$$

Here, $\boldsymbol{e} = [1\ 1\ 1...1]^T$ is a column vector consisting of M unit elements. The choice $h[k] = 1/M$ for $k = 0, 1, ..., (M-1)$, gives the minimum P_t for a low distortion loop. It can also be shown that choosing all tap weights of the decimation filter as $1/M$ minimizes the thermal noise for any loop filter [9].

The estimation process for the contribution of the quantization error to the output is similar to that performed previously for the thermal noise. Assume that the error $e[k]$ behaves as zero-mean noise with uncorrelated samples and that the samples have a mean square value of $\Delta^2/12$, where Δ represents the step size of the quantizer. (Note that this assumption is subject to conditions that ensure the randomness of the error, which may require the use of a dither signal in the loop.) Let $n[k]$ be the impulse response of the quantization NTF from the quantizer to the loop output. It is the inverse transform of the NTF(z) of the loop, windowed by the reset pulse. Then the power of the output quantization noise P_q can be expressed using the triangular $M \times M$ matrix $\boldsymbol{N}$ generated from the $n[k]$ samples in the same way as the $\boldsymbol{S}$ matrix was generated by the $s[k]$ samples. This gives

$$P_q = \frac{\Delta^2}{12} \boldsymbol{h}^T \boldsymbol{N}^T \boldsymbol{N} \boldsymbol{h} \quad (3.27)$$

We want to minimize the output quantization noise power P_q given by Eq. (2.11), subject to constraint (3.26). Using the Lagrange multiplication method [9], this can be achieved analytically. The resulting optimum impulse response for the decimation filter is given by

$$h_{\text{opt}} = \frac{Re}{\boldsymbol{e}^T Re} \quad (3.28)$$

Here, $\boldsymbol{R} = [\boldsymbol{N}^T \boldsymbol{N}]^{-1}$, and $\boldsymbol{e}$ represents the unit-element vector defined above. The matrix $\boldsymbol{N}^T \boldsymbol{N}$ cannot be singular due to the structure of $\boldsymbol{N}$, and thus $\boldsymbol{R}$ must always exist. Software (for example, MATLAB's *quadprog* function) can be used to find h_{opt} from Eq. (3.28). From the IADC model, it can be predicted that for an Lth-order loop the first L elements of h_{opt} should be zero or very small, since the last L output samples of the loop will contain quantization error samples that cannot be canceled by subsequent samples. Therefore, their weight factors $h(0)$, $h(1)$, ... must be small in the optimum solution.

It is also possible to minimize the sum of P_t and P_q, subject to holding the filter's dc gain as 1, by choosing the digital filter impulse response $h(k)$ appropriately. The power of the combined noises at the digital filter output is given by $\boldsymbol{h}^T\boldsymbol{Oh}$, where

$$O = \frac{\gamma k_B T}{C_{in}}S^T S + \frac{\Delta^2}{12}N^T N \qquad (3.29)$$

Consequently, the task becomes to find $h(k)$ so as to achieve

$$\min_h (P_t + P_q) = \min_h (\boldsymbol{h}^T\boldsymbol{Oh}) \qquad (3.30)$$

subject to $\boldsymbol{e} \cdot \boldsymbol{h} = 1$. The process described for minimization of P_q applies also to this general case, and h_{opt} is still given by Eq. (3.28), but now $\boldsymbol{R} = [\boldsymbol{O}^T\boldsymbol{O}]^{-1}$ holds, where $\boldsymbol{O}$ is given by Eq. (3.29). The optimum impulse response $\{h[k]\}$ of the digital filter will then be a compromise between those applicable in the extreme cases described above. Figure 3.9 shows the results obtained using the MATLAB™ code for the example described in Ref. [9] to yield the results shown in Figure 3.9. The code fragment providing this result is given in Ref. [6], Section 12.

The curves indicate the optimal $\{h[k]\}$ responses for minimizing the thermal noise power P_t (dashed line), quantization noise power P_q (dotted curve), and total output noise power (continuous curve). Despite the fact that the areas under the three curves are same due to the dc gain requirement, their individual properties vary, as previously discussed. Thus, the response for minimum thermal

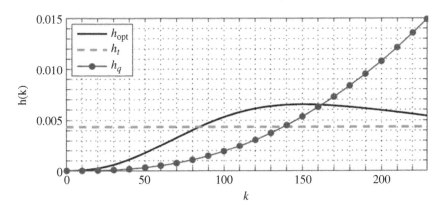

Figure 3.9 Optimal impulse responses of third-order IADC digital filters.

output noise is constant, whereas the response for minimum quantization output noise is similar to a parabola. Optimal total noise is achieved by a response $h(k)$ following initially the quantization noise response, because this suppresses the large noise introduced at the end of conversion. Afterward, the curve follows the thermal noise-minimizing response.

3.3.4 Multiple-Stage IADCs and Extended Counting ADCs

A number of changes can be made to the IADC to improve the SQNR, just as with a $\Delta\Sigma$ ADC: increasing the order L, or the OSR M, or the resolution of the internal quantizer can all improve the SQNR. These measures, however, are all limited by their practical side effects. The OSR M is often restricted to a relatively low value by the amplifiers' bandwidths, or by limits on dissipated power for wideband ADCs. The SQNR cannot be significantly improved by increasing the order of the loop filter for a given OSR, and very high SQNR can often only be obtained by using impractically high quantizer resolutions.

As will be discussed in Chapter 4, a multi-stage noise-shaping (MASH) architecture can be used to solve the problems presented by a low OSR. Using this procedure, the first stage's quantization error e_1 is obtained in an analog form, fed to a second stage, converted to a digital number, and it is canceled by the second stage's output. In a similar manner, the error e_2 in the second stage can be canceled by the output of the third stage, and so on. Then, the digital outputs of all stages are combined using added error-canceling filters H_1, H_2, etc. This allows the designer to obtain high-order noise shaping while using low-order individual loops. Additionally, if the first loop contains a multi-bit quantizer, the error e_1 will be smaller than the full-scale voltage. Hence, when it is fed into the second stage, e_1 may be amplified by a gain $A > 1$, and then a gain $1/A$ can be applied to the output of the second stage. This reduces the final error.

Originally developed for $\Delta\Sigma$ DACs and ADCs, the MASH concept is also applicable to IADCs. Reference [13] described a MASH IADC with two stages, where the first stage operates from clock period 1 to period M, while the second stage runs continuously from the second clock period until the clock period $(M + 1)$. In fact, many stages of IADC can be cascaded. Reference [3] describes an eight-stage 12-bit IADC operating with an OSR of just three!

One can obtain an economic configuration for a MASH IADC by observing that the scaled last quantization error $e_1[M]$ generated in the first loop gives a delayed version of the total conversion error of the first loop. In general, $e_1[M]$ can be calculated by subtracting the input of the first-stage quantizer from its output. However, this is complicated. But for the low-distortion structure with a maximally flat NTF(z), Eq. (3.9) shows that $e[M]$ can be determined simply from the output $x_3[M]$ of the last integrator in the first loop. In an efficient MASH IADC, a second stage would then be inactive until the clock period $(M - 1)$,

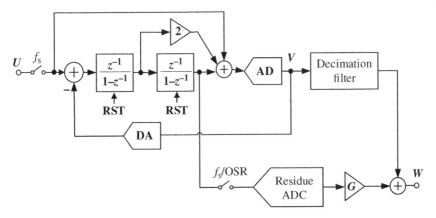

Figure 3.10 An extended-counting ADC of 2–0. Source: Adapted from Rombouts et al. [15].

and then it would convert and scale $x_3[M]$, while the output of the first stage is being processed by the decimation filter, creating the most significant bits of the output word. In the second stage, the N_{LSB} least significant bits of the output word will be produced. A Nyquist-rate ADC can be used for the second stage (e.g. a successive-approximation ADC), and pipelined operations can be realized if $N_{LSB} < (M - 1)$. Such multistage IADCs are often called *extended-range* or *extended-counting* ADC [14–17]. They can be denoted as an L_1–L_2 MASH IADC, where L_1 and L_2 are the orders of the first and second stages, respectively. As an example, Figure 2.27 shows a 2–0 MASH IADC [17]. In the first stage, a low-distortion second-order IADC was used, and in the second stage, a SAR ADC. In a bandwidth of 0.5 MHz, it achieved SNDR >86 dB (Figure 3.10).

3.4 Estimation of Power Consumption

Power consumption is one of the most important performance parameters. Generally, the accuracy of high-solution ADCs is limited by the sampled kT/C thermal noise, which determines the capacitor size. The power consumption is largely determined by the amplifier transconductances needed to drive the capacitive loads, including large signal slewing and small signal settling. In this section, the power dissipation in switched-capacitor IADCs is discussed and estimated.

3.4.1 Power Consumption for Small Signal Settling

In a switched-capacitor integrator with a two-phase nonoverlapping clock of a frequency $f_s = 1/T_s$, the circuit performs sampling in the first half clock period

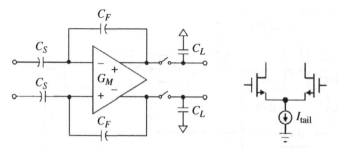

Figure 3.11 A switched-capacitor integrator circuit in integration phase.

and integration in the other half clock period. In the integrator shown in Figure 3.11, the opamp's transconductance G_M must be designed such that the integrator's dynamic transient response is able to perform slewing and small signal settling within an acceptable error in half clock period ($T_s/2$). To make the power estimation simple, the opamp in the power estimation is assumed to be a simple differential pair biased by a tail current I_{tail}. The feedback factor β, the total capacitive load $C_{L,eq}$, and the gain-bandwidth product (GBW) f_0 are denoted as

$$\beta = \frac{C_F}{C_S + C_F}; \quad C_{L,eq} = \frac{C_S \cdot C_F}{C_S + C_F} = \beta \cdot C_S; \quad \omega_0 = 2\pi f_0 = \frac{G_M}{C_{L,eq}} \quad (3.31)$$

The opamps for multistage IADCs usually need not have very high gain, and two-stage opamps might be adequate.

A duty ratio d of the half clock period may be assigned to small signal settling, where $0 < d\% < 100$, and 50% is a usual choice. The opamp's time constant $\tau = 1/(\beta \cdot \omega_0)$ is designed to achieve a settling error within 0.1% error within the given time period $d \cdot T_s/2$. This requires

$$e^{-\beta \cdot \omega_0 \cdot d \cdot T_s/2} < 0.1\% \quad (3.32)$$

$$d \cdot \frac{T_s}{2} \geq \frac{7}{\beta \cdot \omega_0}; \quad \beta \cdot \omega_0 = \frac{\beta \cdot G_M}{C_{L,eq}} = \frac{G_M}{C_S} \quad (3.33)$$

Defining the signal bandwidth as f_{BW}, the oversampling frequency can be described as $f_s = 2 \cdot \text{OSR} \cdot f_{BW}$. By Eqs. (3.31)–(3.33), the opamp's G_M can be estimated from

$$G_M \geq C_S \cdot 14\frac{f_s}{d}$$

$$G_M \geq \text{OSR} \cdot C_S \cdot 28 \cdot f_{BW}/d \quad (3.34)$$

In linear transistor operation, the transconductance is determined by the overdrive voltage $V_{OV} \equiv V_{GS} - V_{TH}$. Then $G_M = 2 \cdot (I_{tail}/2)/V_{OV}$. Hence, the bias current of the differential pair's tail current I_{tail} must satisfy

$$I_{tail} \geq V_{OV} \cdot OSR \cdot C_S \cdot 28 \cdot f_{BW}/d \tag{3.35}$$

As described in Section 3.3.2, higher-order IADCs suffer from higher kT/C noise due to higher noise penalty factor N_f. The sampled kT/C noise gives the bound of SNR [6]

$$SNR = 10 \cdot \log \frac{V_{FS}^2/8}{5 \cdot N_f \cdot \frac{kT}{C_S} \cdot \frac{1}{OSR}} \tag{3.36}$$

In Eq. (3.35), for a two-phase clock, the noise power kT/C is injected twice. Fully differential circuits double the noise power again. Including the opamp thermal noise, a total $5kT/C$ can be used to estimate the sampled thermal noise power. The flicker noise is not taken into account here, and it must be simulated and accounted in finding the noise power, especially for low signal bandwidth. The sampling capacitor size can be determined by the SNR requirement. For a given OSR

$$C_S \cdot OSR \geq \frac{5kT \cdot N_f \cdot 10^{(SNR/10)}}{V_{FS}^2/8} \tag{3.37}$$

The current consumption can, therefore, be estimated as

$$I_{tail} \geq V_{OV} \frac{5kT \cdot N_f \cdot 10^{(SNR/10)}}{V_{FS}^2/8} \cdot \frac{28 \cdot f_{BW}}{d} \tag{3.38}$$

The sizes of the sampling capacitors C_S for the second and third integrators in a higher-order IADC can be scaled down when calculating the total noise referred to IADC's input. The bias current of later stages can be scaled down accordingly.

3.4.2 Power Consumption for Slewing

Switched-capacitor circuits are sampled-data system, and usually, the step response causes large signals in the transient response. A transistor circuit in large signal operation will need to slew to recover back to small signal operation. In a feedforward modulator with a direct input path, for example, in Figure 3.3, the integrator output step response V_{step} is about the size of the internal quantizer's LSB voltage $V_{FS}/(L-1)$ [6]. When the time $(1-d) \cdot (T_S/2)$ is given for large signal slewing, the biasing current for large signal operation can be estimated from

$$\frac{I_{tail}/2}{C_{L,eq}} \geq \frac{V_{step}}{(1-d) \cdot (T_S/2)} \tag{3.39}$$

Taking into account the bandwidth, the voltage step size, and the load capacitance, the biasing current can be estimated from

$$I_{\text{tail}} \geq \frac{8 \cdot \text{OSR} \cdot C_{L,\text{eq}} \cdot f_{\text{BW}}}{(1 - d)} \cdot \frac{V_{\text{FS}}}{L - 1} \tag{3.40}$$

Since the slewing may dominate the dynamic settling response, the slew-reduction property of the feedforward multilevel modulator due to its suppression of the signal in the loop filter is of great usefulness.

3.4.3 An Example

The specification of a single-loop IADC is as follows:

SNR = 96 dB, V_{FS} = 2 V, VDD = 1.8 V, 25 kHz bandwidth, OSR = 16.

As Figure 3.5 and Eq. (3.20) show, the very low OSR will require a fourth-order modulator with 9-level internal quantizer to achieve an SNR of 96 dB. The IADC2 circuit in Figure 3.3 can be extended to fourth order by adding two more integrators and changing the resolution of the internal quantizer to nine levels. Also, the fourth-order IADC suffers from a higher kT/C noise penalty, which is twice that of a conventional $\Delta\Sigma$ ADCs, as described in Section 3.3.2. Taking into account the noise penalty $N_f = 2$, Eq. (3.38) must be changed to

$$I_{\text{tail}} \geq V_{\text{OV}} \frac{10kT \cdot 10^{(\text{SNR}/10)}}{V_{\text{FS}}^2/8} \cdot \frac{28 \cdot f_{\text{BW}}}{d} \tag{3.41}$$

We choose the overdrive voltage as $V_{\text{OV}} = 0.1$ V and assign the same duty cycle of both small signal settling and slewing as $d = 50\%$.

The power consumption for small signal settling can be estimated as follows:

1. The current and power consumptions of the first integrator (INT1) can be calculated as 46.15 µA and 83.1 µW, respectively, from Eq. (3.41).
2. When referred to the IADC's input, the kT/C noise powers of the second, third, and fourth integrators are attenuated by the noise-shaping gains $(\text{OSR}/\pi)^2$, $(\text{OSR}/\pi)^4$, and $(\text{OSR}/\pi)^2$, respectively [6]. To consider both the noise scaling and the capacitor matching requirements, it is a good choice to scale down the second integrator's sampling capacitor size only by (OSR/π), instead of $(\text{OSR}/\pi)^2$. For the second, third, and fourth integrators, the sampling capacitor is scaled down as $C_S/(\text{OSR}/\pi)$, $C_S/(\text{OSR}/\pi)/2$, and $C_S/(\text{OSR}/\pi)/2$, respectively. Since the OSR/$\pi \sim 5$, the current consumption needed to drive the capacitive loads can be scaled down as 83.1 µW $\cdot$ (1 + 0.2 + 0.1 + 0.1) = 116.3 µW.
3. The digital power consumption, including the two-phase nonoverlapping clock and drivers, the feedback DAC calibration circuits, and all other logic circuits, can be estimated to be about the same as analog power.
4. Thus, the total estimated power is 232.6 µW.

The power consumption for large signal slewing can be estimated as follows:

1. Fifty percent duty of half clock period is assigned for slewing. Since the C_S must be doubled due to the noise penalty, the C_S is calculated as 20.6 pF based on Eq. (3.37). The INT1's coefficient is 1; this leads to feedback factor $\beta = 0.5$, and a load capacitance $C_{L,eq} = 10.3$ pF. Equation (3.40) can be used to calculate the bias current of first integrator. It is 16.48 µA.

2. The first integrator's bias current must satisfy both the large signal slewing and small signal settling. In this case, a nine-level quantizer is used and thus the integrator swing is reduced significantly. When 46.15 µA is assigned as the bias current, it fulfills both the requirements.

3. Usually, the slewing in switched-capacitor circuits needs more time or more bias current. The estimation given above may be too optimistic and the realized power consumption higher.

Another limitation which was not taken into account is the harmonic distortion. If a 3 dB margin is considered for the distortion, the IADC will end up with SNDR = 93 dB. The Schreier (FoM_S) and Walden FoM (FoM_W) figures of merit are commonly used to evaluate the performance of ADCs. The parameters of the ADC considered above can be summarized as shown below:

VDD (V)	V_{FS} (V)	SNR (dB)	SNDR (dB)	OSR	INT1 C_S (pF)	INT1 power (µW)	Total power (µW)	FoM_W (fJ/conv.)	FoM_S (dB)
1.8	2	96	93	16	20.6	83.1	232.6	127.4	173.3

$\text{FoM}_S = \text{SNDR} + 10 \cdot \log(f_{BW}/\text{power})$
$\text{FoM}_W = \text{power}/(2^{ENOB} \cdot 2f_{BW})$

References

1 J. Markus, J. Silva, and G. C. Temes, "Theory and applications of incremental delta sigma converters," *IEEE Transactions on Circuits and Systems I: Regular Papers*, vol. 51, no. 4, pp. 678–690, Apr. 2004.

2 C.-H. Chen, Y. Zhang, T. He, P. Chiang and G. C. Temes, "A micropower two-step incremental analog-to-digital converter," *IEEE Journal of Solid-State Circuits*, vol. 50, no. 8, pp. 1796–1808, Aug. 2015.

3 T. C. Caldwell and D. A. Johns, "Incremental data converters at low oversampling ratios," *IEEE Transactions on Circuits and Systems I: Regular Papers*, vol. 57, no. 7, pp. 1525–1537, Jul. 2010.

4 R. Schreier. Delta Sigma Toolbox (https://www.mathworks.com/matlabcentral/fileexchange/19-delta-sigma-toolbox), MATLAB Central File Exchange, 2022.

5 S. Gambini, K. Skucha, P. P. Liu, J. Kim, and R. Krigel, "A 10 kPixel CMOS hall sensor array with baseline suppression and parallel readout for immunoassays," *IEEE Journal of Solid-State Circuits*, vol. 48, no. 1, pp. 302–317, Jan. 2013.

6 S. Pavan, R. Schreier, and G. C. Temes, *Understanding Delta-Sigma Data Converters*. Pascataway, NJ, USA: IEEE Press/Wiley, 2017.

7 S. Kavusi, H. Kakavand and A. El Gamal, "On incremental sigma-delta modulation with optimal filtering," *IEEE Transactions on Circuits and System I: Regular Papers*, vol. 53, no. 5, pp. 1004–1015, May 2006.

8 O. Oliaei, "Sigma-delta modulator with spectrally shaped feedback," *IEEE Transactions on Circuits and Systems II*, vol. 50, no. 9, pp. 518–530, Sept. 2003.

9 Y. Zhang, C.-H Chen, T. He, P. Chiang and G. C. Temes, "A 16b multi-step incremental analog-to-digital converter with single-opamp multi-slope extended counting," *IEEE Journal of Solid-State Circuits*, vol. 52, no. 4, pp. 1066–1076, Apr. 2017.

10 S. -C. Kuo, J. -S. Huang, Y. -C. Huang, C. -W. Kao, C. -W. Hsu and C. -H. Chen, "A Multi-Step Incremental Analog-to-Digital Converter With a Single Opamp and Two-Capacitor SAR Extended Counting," in *IEEE Transactions on Circuits and Systems I: Regular Papers*, vol. 68, no. 7, pp. 2890–2899, Jul. 2021.

11 E. B. Hogenauer, "An economical class of digital filters for decimation and interpolation," *IEEE Transactions on Acoustics, Speech Signal Processing*, vol. 29, pp. 152–159, Apr. 1981.

12 J. Steensggard et al., "Noise-power optimization of incremental data converters," *IEEE Transactions on Circuits and Systems I: Regular Papers*, vol. 55, no. 5, pp. 1289–1296, Jun. 2008.

13 J. Robert and P. Deval, "A second-order high resolution incremental A/D converter with offset and charge injection compensation," *IEEE Journal of Solid-State Circuits*, vol. 23, no. 3, pp. 736–741, Mar. 1988.

14 R. Harjani and T. A. Lee, "FRC: a method for extending the resolution of Nyquist-rate converters using oversampling," *IEEE Transactions on Circuits and Systems-II*, vol. 45, no. 4, pp. 482–494, Apr. 1998.

15 P. Rombouts, W. de Wilde, and L. Weyten, "A 13.5-b 1.2-V micropower extended counting A/D converter," *IEEE Journal of Solid-State Circuits*, vol. 36, no. 2, pp. 176–183, Feb. 2001.

16 J. De Maeyer, P. Rombouts, and L. Weyten, "A double-sampling extended-counting ADC," *IEEE Journal of Solid-State Circuits*, vol. 39, pp. 411–418, Mar. 2004.

17 A. Agah, K. Vleugels, P. B. Griffin, M. Ronaghi, J. D. Plummer, and B. A. Wooley, "A high-resolution low-power incremental $\Sigma\Delta$ ADC with extended range for biosensor arrays," *IEEE Journal of Solid-State Circuits*, vol. 45, pp. 1099–1110, Jun. 2010.

4

Multistage and Extended Counting Incremental Analog-to-Digital Converters

A first-order incremental analog-to-digital converter (IADC1) needs 2^N oversampling clock periods for N-bit accuracy, requiring a high sampling frequency. It is also usually not energy-efficient. To enhance energy efficiency, higher-order modulators are preferred to increase the resolution within the same conversion time. However, single-loop high-order modulators are prone to instability and have a reduced non-overloading input range. As an alternative to single-loop modulators, multistage noise shaping (MASH) IADCs [1, 2] may be used. Also, hybrid schemes, which incorporate an added Nyquist-rate ADC to perform extended counting, are excellent candidates for high-resolution IADCs.

In this chapter, the design and operation of MASH IADCs with a feedforward modulator [3] will be described in Section 4.1. Hybrid IADCs performing extended counting, and thus achieving excellent energy efficiency, are discussed in Section 4.2. A design example, which describes the use of hardware-sharing two-capacitor extended counting, is described in Section 4.3. Section 4.4 sums up the material provided in this chapter, and draws the conclusions.

4.1 Multistage Noise Shaping (MASH) Incremental ADCs

4.1.1 The Design of MASH IADCs

An effective scheme for mitigating the stability problem of conventional higher-order single-loop $\Delta\Sigma$ modulators and allowing a wide non-overloaded range is the MASH technique [1, 2]. Figure 4.1 shows a conventional MASH 1-1 delta-sigma modulator, in which two first-order loops are cascaded to accomplish second-order noise shaping, with additional error cancellation logic (ECL) used to perform a filtering by $(1 - z^{-1})$ in the digital domain. The decimation filter can be implemented by cascading a conventional Hogenauer sinc filter and half-band filters [1].

Incremental Data Converters for Sensor Interfaces, First Edition. Chia-Hung Chen and Gabor C. Temes.
© 2024 The Institute of Electrical and Electronics Engineers, Inc. Published 2024 by John Wiley & Sons, Inc.

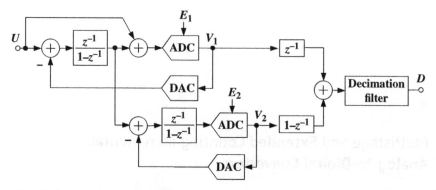

Figure 4.1 A 1-1 MASH modulator.

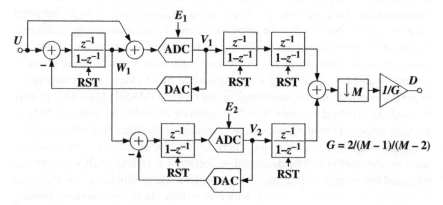

Figure 4.2 A 1-1 MASH IADC.

The MASH technique can also be applied to a higher-order IADC by cascading lower-order IADC modulators [3, 4] with wide non-overloaded range. Figure 4.2 shows an example of 1-1 MASH second-order IADC2 obtained by cascading two IADC1s. At the end of data conversion (time index $i = M$), the closed-loop difference equation in the time domain is

$$W_1[M] = \sum_{i=1}^{M-1} (U[i] - V[i]) \tag{4.1}$$

$$\sum_{k=1}^{M-1}\sum_{i=1}^{k-1} (U[i] - V_1[i]) - \sum_{i=1}^{M-1} (V_2[i]) + E_2[M] = V_2 \tag{4.2}$$

A constant input voltage $U[i]$ twice integrated would give a value $U[i](M-1)/(M-2)/2$. Hence, defining the average input voltage $\widetilde{U}$ as in (4.3), it can be reconstructed from the output bit streams V_1 and V_2. Equation (4.4) describes the result.

Only simple cascaded counters are needed to reconstruct the input according to the right-hand side of Eq. (4.4). The quantization error E_1 of the first loop is canceled, and it does not appear in the MASH output. The error E_2 of the second loop is scaled by $2/(M-1)(M-2)$; it is second-order noise shaping.

$$\tilde{U} = \frac{2}{(M-1)(M-2)} \sum_{k=1}^{M} \sum_{i=1}^{k} U[i] \tag{4.3}$$

$$\tilde{U} + \frac{2}{(M-1)(M-2)} E_2[M] = \frac{2}{(M-1)(M-2)} \left\{ \sum_{k=1}^{M-1} \sum_{i=1}^{k-1} V_1[i] - \sum_{i=1}^{M-1} V_2[i] \right\} \tag{4.4}$$

The 1-1 MASH IADC can be extended to a third-order 1-1-1 MASH, which is equivalent to a third-order IADC. The design Eqs. (4.1)–(4.4) can be extended to the design of the matched decimation filter used to reconstruct the input signal. Alternatively, the first loop can be a second-order loop filter and a 2-1 MASH can be implemented as a third-order IADC. An example of 2-1 MASH IADC is illustrated in Figure 4.3. The analysis of a 2-1 MASH IADC example is derived in Chapter 5.

4.1.2 Trade-Offs in the Design of MASH IADCs

Conventional MASH $\Delta\Sigma$ modulators need to use error canceling logic circuits before adding the bit streams of the individual loops, in order to cancel the quantization errors of the most significant bit (MSB) loops. The opamp dc gains in the

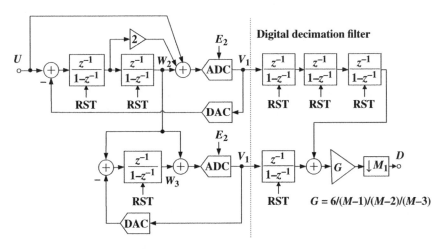

Figure 4.3 A 2-1 MASH incremental ADC.

first loop need to be very high, to avoid signal-to-quantization-noise ratio (SQNR) degradation caused by mismatch between the analog and digital realizations of the noise transfer function containing powers of $(1 - z^{-1})$. Thus, a MASH $\Delta\Sigma$ ADC for a 16-bit signal-to-noise ratio (SNR) usually requires opamp dc gains of at least 90 dB [2], which is very difficult to achieve in a low-voltage design. In MASH IADCs [3, 4], the oversampled bit streams of the $\Delta\Sigma$ loops are accumulated in one or more cascaded counters. The decimation filter is much simpler than those needed for conventional $\Delta\Sigma$ modulators. The Nyquist-rate data from each loop are accumulated separately, and the counters dump the data before the next reset pulse. The requirements of the opamp gain and coefficient matching in a MASH IADC are, hence, much more relaxed than in their conventional MASH $\Delta\Sigma$ counterparts.

4.1.3 Hybrid Schemes for an IADC and a Nyquist-Rate ADC

In the scheme of Figure 4.1, the feedforward loop filter processes only the shaped quantization noise. At the end of the conversion, the voltage stored at the output of the last integrator is the residue voltage of the $\Delta\Sigma$ data conversion, and it is available for fine quantization [5–10]. The residue voltage acquisition is illustrated in Figure 4.4a. Instead of using higher-order IADCs, it is possible to use a combination of a lower-order IADC and a Nyquist-rate ADC. Then, an energy-efficient successive approximation register (SAR) or cyclic ADC operated at Nyquist rate can sample the residue voltage right before the reset pulse, and perform the fine quantization. Figure 4.4b illustrates an extended counting circuit [6] using an IADC2 to oversample and coarse-quantize the input signal. When the modulator is running in a continuous mode without reset, a z-domain analysis gives $W_2(z) = -z^{-2} \cdot E_1(z)$, corresponding to a delay of two oversampling clock periods. When the reset signal terminates the operation at the end of IADC2's conversion period (time index $i = M$), the second integrator output $W_2[M] = \sum_{k=1}^{M-1} \sum_{i=1}^{k-1}(U[i] - V[i])$ delivers the residue voltage to the second ADC, a SAR circuit. Afterward, the SAR ADC fine-quantizes the residue voltage at the Nyquist rate, resulting in the digital signal D_2:

$$\sum_{K=1}^{M-1}\sum_{i=1}^{K-1}(U[i] - V[i]) + E_2 = D_2 \tag{4.5}$$

Using Eq. (4.3) to define the average input voltage $\widetilde{U}$, the input signal can be reconstructed from the output bit streams $V[i]$ and D_2

$$\widetilde{U} + \frac{2}{(M-1)(M-2)}E_2 = \frac{2}{(M-1)(M-2)}\left(\sum_{K=1}^{M-1}\sum_{i=1}^{K-1}V[i] + D_2\right) \tag{4.6}$$

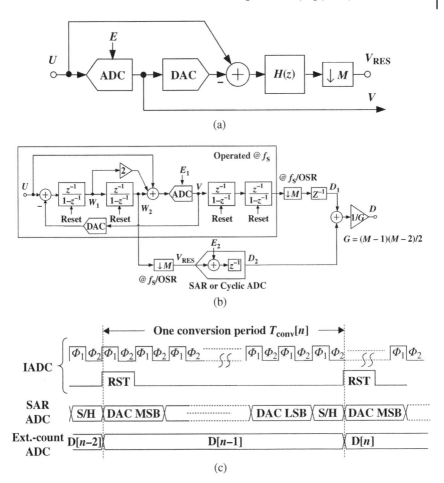

Figure 4.4 (a) Residue voltage acquisition. (b) An IADC using a Nyquist-rate ADC for extended counting. Source: Adapted from Agah et al. [6]. (c) The simplified timing diagram for one conversion period.

Thus, the quantization error E_1 of the IADC2 loop is canceled, and only E_2 appears after combination of the data streams. If the second ADC has a 10-bit resolution, the SQNR of this hybrid ADC is equivalent to an IADC with a 10-bit quantizer. With proper design of the digital summation logic and decimation filter, the two cascaded loops can achieve a very high resolution with outstanding energy efficiency. However, the last integrator in the first stage IADC needs to drive the large input capacitance of the 11-bit SAR, and therefore, needs additional power.

4.1.4 Extended Counting with Hardware Sharing

For high-resolution and narrow signal bands, the time required for data conversion is usually long. Hence, instead of cascading two loops, the conversion can be performed in two steps, and the hardware used in the two steps can be shared to improve energy efficiency. An example of a hardware-sharing extended-counting scheme is shown in Figure 4.5a [7, 10]. A discrete-time IADC1 performs the coarse quantization (Figure 4.5b), and the integrator stores the residue voltage at the end of the first quantization step. In the second step, the hardware is reused and reconfigured as a 10-bit cyclic ADC to continue with the fine quantization (Figure 4.5c). By sharing the hardware, the energy efficiency is improved significantly.

Reviewing the SQNR variation with oversampling ratio (OSR) shows that every doubling of the OSR will improve the SQNRs of an IADC1 and IADC2 by

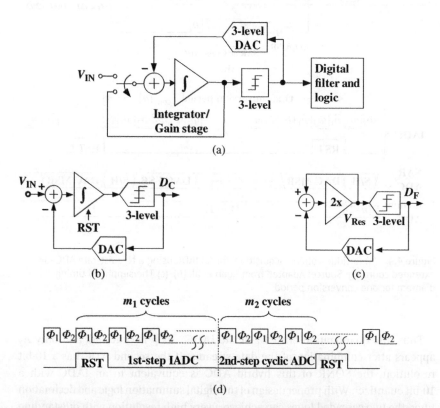

Figure 4.5 (a) An example of IADC with extended counting using hardware sharing. Source: Kim et al. [7], Katayama et al. [10]. (b) A discrete-time IADC1 acts as the coarse quantization ADC. (c) Reconfigured as a 10-bit cyclic ADC to perform the fine quantization. (d) Simplified timing diagram for two-step operation.

approximately 6 and 12 dB, respectively. However, binary search in a Nyquist-rate ADC can improve by 6 dB in every clock period. A more energy-efficient scheme for high-resolution data conversion is, hence, to accomplish one conversion by both oversampling and binary searching in multistep operations: using an IADC to perform the coarse quantization, and then reconfiguring the hardware to perform a binary search on the residue voltage for the fine quantization [7–9, 11, 12]. In Figure 4.6a, the SQNRs and IADC1 with and without extra binary extended counting are plotted along with those of the single-loop IADC1 and IADC2. Each extra clock period boosts up the SQNR by 6 dB during extended counting. Figure 4.6b plots the SQNR during clock indices from the 250th to 270th clock periods. A cyclic ADC is used to perform the binary search.

In Ref. [10], an IADC1 with 31-level internal quantizer is described. It performs coarse quantization for 32 oversampled clock periods, and then it is reconfigured as a cyclic ADC for another 8 clock periods. By sharing the opamp during the operation of two steps, this hybrid ADC achieves an excellent Schreier figure-of-merit of 173.6 dB. However, reconfiguring the IADC into the cyclic ADC with a multiplying gain of 2.25 complicates the overall design. Such complicated configuration usually suffers more from parasitics, introduces larger mismatch, and gains errors between stages. The required component matching requires additional circuitry. Alternatively, calibration circuitry may be incorporated to mitigate the SQNR degradation caused by the mismatches or by stage gain errors.

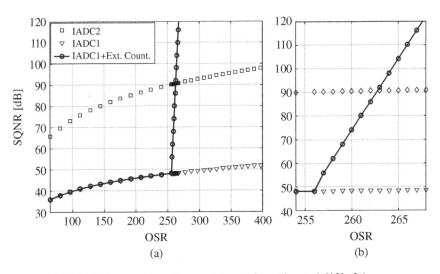

Figure 4.6 (a) SQNR comparison. Source: Adapted from Kim et al. [10] of the binary-extended counting scheme with single-loop IADC1 and IADC2. (b) After using the IADC1 for 256 clock periods, the SQNR is boosted by 6 dB/clock period during the extended counting.

4.2 Design Examples

4.2.1 IADC with Two-Capacitor Counting

A two-step IADC with two capacitors binary-extended counting which retains the advantages of extended counting and simplifies the hardware reconfiguration is described next.

Figure 4.7a shows the simplified scheme of a serial digital-to-analog converter (DAC) [13]. Capacitor values C_1 and C_2 are nominally equal. Initially, the capacitor voltages V_1 and V_2 are reset. During the first half phase (Φ_1), C_1 is pre-charged either to the reference voltage V_{REF} if the least significant input bit $D = 1$, or to ground if $D = 0$. In the next phase (Φ_2), the charges of the two capacitors are shared. The process is repeated for all bits, ending with the MSB. A digital input sequence $D = 1101$ results in $V_2/V_{REF} = 13/16$. The transient waveform is illustrated in Figure 4.7b.

With the addition of a comparator, shift registers, and sequencing logic, the serial DAC can be used to construct a SAR ADC [13]. However, challenging nonidealities limit the accuracy, linearity, and output level. The top parasitic capacitances directly introduce capacitor mismatch errors. The asymmetry of the switches' nonlinear source/drain junction capacitances results in a net feedthrough error voltage during charge redistribution. Error cancellation and large sizes of C_1 and C_2 may still not be sufficient to achieve good SNR.

In spite of the drawbacks of the 2-C DAC, this simple scheme can be useful in a hybrid ADC to perform fine quantization [11, 12, 14, 15]. Figure 4.8 depicts the equivalent z-domain model, with the simplified timing control and timing generator shown in Figure 4.9. The conversion period T_{CONV} between

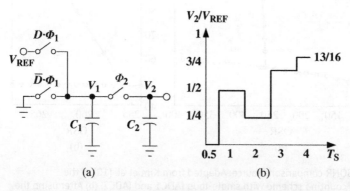

Figure 4.7 (a) Serial charge-redistribution DAC. Source: Adapted from Suarez et al. [13]. (b) Illustration of DAC signal sequences for an input word $D = 1101$.

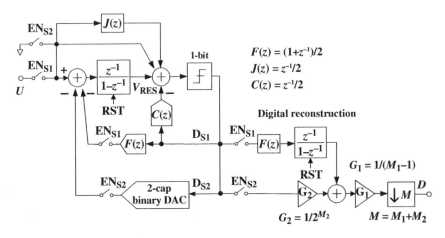

Figure 4.8 System-level z-domain block diagram of the IADC1 with SAR binary-extended counting.

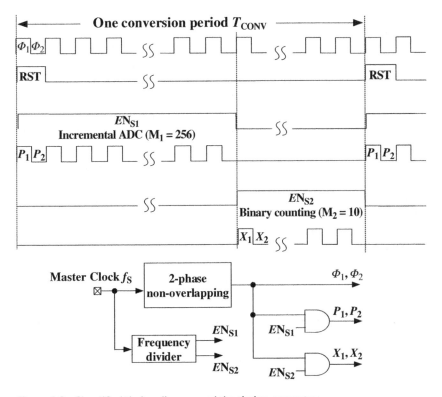

Figure 4.9 Simplified timing diagram and the timing generator.

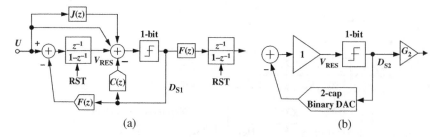

(a)   (b)

Figure 4.10 (a) Equivalent block diagram during step 1: IADC1 with an FIR feedback DAC. (b) Equivalent block diagram during step 2: the 2-C SAR ADC performing the extended counting.

two adjacent reset signals is divided into two-time intervals: the first interval comprised of 256 clock cycles ($M_1 = 256$), and the second one of 10 clock cycles ($M_2 = 10$). As shown in Figure 4.10a, the circuits in the first step realize an IADC1 with a one-bit internal quantizer and a feedback finite impulse response (FIR) DAC. The two-tap FIR path $F(z) = (1 + z^{-1})/2$ is incorporated to reduce the integrator's output step sizes. The reduced voltage step sizes help to relax the slew rate requirements significantly during transient settling, and the bias current and power consumption of the opamp are, hence, reduced [16, 17]. The added delay in the loop must be compensated, and it needs another path $C(z) = \frac{z^{-1}}{2}$ to restore the original noise transfer function. In addition, a path $J(z) = z^{-1}/2$ is added in order to restore the signal transfer function (STF) to unity, and to allow the loop filter to process the quantization noise only. The quantization residue $V_{RES}[M_1] = \sum_{i=1}^{M_1-1} U[i] - \sum_{i=1}^{M_1-1} D_{S1}[i]$ is accumulated at the integrator's output after M_1 clock periods, where D_{S1} is the digital output of IADC1.

The second step starts when EN_{S2} goes high. Reusing the integrator as a buffer and a 2-C DAC, as shown in Figure 4.10b, the quantization residue V_{RES} is quantized by a fine binary search SAR ADC during the next M_2 clock periods. The result is

$$V_{RES}[M_1 + M_2] + E_2 = \sum_{i=1}^{M_2} 2^{M_2-i} \cdot D_{S2}[i] \tag{4.7}$$

here, E_2 is the quantization error of the SAR during the fine quantization. Equation (4.7) can also be written in the form

$$V_{RES}[M_1 + M_2] = \sum_{i=1}^{M_1-1} U[i] - \left[\sum_{i=1}^{M_1-1} D_{S1}[i] - \frac{1}{2^{M_2}} \sum_{i=1}^{M_2} 2^{M_2-i} \cdot D_{S2}[i] \right] \cdot V_{REF}$$

$$\tag{4.8}$$

When the SAR finished the fine quantization after $M_1 + M_2$ clock periods, the conversion cycle is finished. $\widetilde{U}$ is defined as the average of the input samples, that is as $\sum_{i=1}^{M_1-1} U[i]/(M-1)$. It is given by

$$\widetilde{U} = \frac{1}{M_1 - 1}\left[\sum_{i=1}^{M_1-1} D_{S1}[i] + \frac{1}{2^{M_2}}\sum_{i=1}^{M_2} 2^{M_2-i} \cdot D_{S2}[i]\right] \cdot V_{FS} + \frac{V_{FS}}{(M_1 - 1) \times 2^{M_2}}$$

(4.9)

As Eq. (4.9) shows, the digitized bit streams D_{S1} and D_{S2} from the two steps need to be added with proper scale factors. The reconstruction circuitry can be designed according to Eq. (4.9). The ideal SQNR of the proposed ADC can be estimated from

$$\text{SQNR} \approx 20\log_{10}\left(M_1 \cdot 2^{M_2}\right) = 20\log_{10}(M_1) + 6.02 \cdot M_2 \tag{4.10}$$

The overall two-step ADC is equivalent to an IADC1 with an internal 10-bit quantizer, in which the IADC1 of OSR = 256 contributes an 8-bit resolution, and the 10-bit binary search contributes an additional 10 bits. The overall effective number of bits is 18 bits. The capacitor matching is usually limited to 10-bit accuracy, and a good choice for the SAR binary search resolution is thus 10 bits. In a prior state-of-the-art IADC [6], the residue voltage is transmitted to the added stage to enhance the resolution. The proposed IADC stores the residue voltage and feeds it into the quantizer. It is, hence, more robust than prior arts to perform extended counting.

4.2.2 Switched-Capacitor Implementation of the Example IADC

For the first step, the equivalent single-ended switched-capacitor circuit is shown in Figure 4.11a. (The actual implementation is fully differential.) Here, the whole circuit performs as a feedforward IADC1. Bootstrapped switches are utilized for sampling to improve the linearity. To fulfill the kT/C noise requirement, theintegrator's input sampling capacitor C_S, and integration capacitor C_F are both sized as 3.2 pF. To implement the FIR path $F(z) = (1+z^{-1})/2$, the sampling capacitor is split into two $C_{DAC1} = 0.5\ C_F$ capacitors. The compensation function $C(z) = z^{-1}/2$ is realized by C_4 directly at the quantizer. The path $J(z) = \frac{z^{-1}}{2}$ realizes the input feedforward coupling. It is implemented by interleaving the samples on the two capacitors C_2 and C_3 [15]. While C_2 samples are the latest input, the previous sample held in C_3 is converted. P_{1o} and P_{1e} alternately control the four switches. The passive switched-capacitor adder using C_0–C_4 carries out the summation at the quantizer's input. C_0 and C_1 are sized as 500 fF, and C_2, C_3, and C_4 are sized as 250 fF. This allows accurate realization of the coefficient ratios. Aggressive design can choose the capacitances much smaller, but this does not save much power and the FoM is not improved significantly.

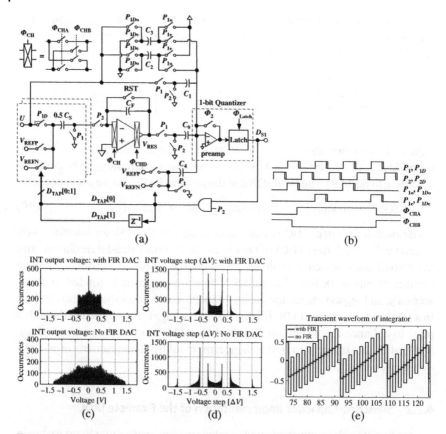

Figure 4.11 (a) Step 1: single-ended switched-capacitor circuit implementation of an IADC1 with FIR feedback DAC. C_S and C_F are sized as 3.2 pF. (b) Additional timing control for input FIR path and chopper. (c) Histogram of the integrator's differential output voltage with FIR DAC (top) versus no FIR (bottom). (d) Histogram of the integrator's differential voltage step. (e) Simulated waveform of the integrator's transient output.

In the digital filters of Figures 4.8 and 4.10, a digitally implemented function $F(z)$ must be added to compensate for the analog FIR $F(z)$. Mismatch between analog $F(z)$ and digital $F(z)$ will cause SQNR degradation. Simulations show that a 0.3% mismatch between the two divided C_{DAC1} values causes a 2 dB SQNR degradation. Since the value of $C_{DAC1} = C_{DAC2} = 1.6$ pF is large enough to allow excellent matching, this degradation is not significant. Using more than two taps in the FIR filter $F(z)$ would bring only marginal benefits, and it would increase the complexity of the input feedforward compensation path $J(z)$, which may affect the overall performance of extended counting.

To suppress the flicker noise and reduce the dc offset, the input virtual ground and the differential output of the opamp are chopped. The chopping frequency is chosen to be $f_s/4$. Chopping occurs at the sampling phase Φ_1 to avoid interference with integration. Figure 4.11b illustrates the additional timing for input FIR path and chopper. The FIR feedback DAC effectively reduces the transient voltage step size. Figure 4.11c plots the histogram and Figure 4.11d shows the transient waveform to compare the difference due to the FIR filter.

The quantization residue V_{RES} stored by the integrator is available for finer quantization. This is carried out in the *second step*. Two capacitors C_R are pre-charged to the reference voltage V_{REFP} and V_{REFN} at the end of the first step and serve as charge reservoirs. The charge reservoirs share the charge with the C_S capacitors and generate the binary-weighted DAC voltage sequentially during the 2-C stage operation. This 2-C DAC configuration and the use of charge sharing avoid wide value spread in the binary-weighted capacitor array. In the actual implementation shown in Figure 4.12, the charge reservoir capacitance 1.6 pF is

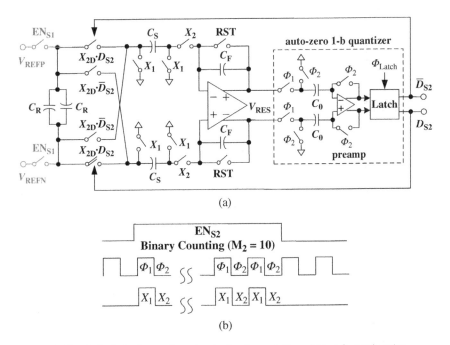

(a)

(b)

Figure 4.12 (a) Switched-capacitor circuits implementation of Step 2: reusing the opamp and 2-C binary DAC as a SAR ADC. Each C_R is 0.8 pF. (b) Simplified timing when binary-extended counting is enabled. The two-phase nonoverlapping clocks Φ_1/Φ_2 and X_1/X_2 are generated as in Figure 4.9.

split into two halves $C_R = 0.8$ pF. By cross-connecting the two halves, the parasitic capacitances on the two terminals can be approximately equalized. Using the 2-C DAC and reusing the same integrator as a buffer to subtract the binary DAC voltage, the overall circuit is now working as a 2-C SAR ADC.

In the second step, when phase X_1 is high, the pre-charged C_Rs are disconnected from V_{REFP} and V_{REFN}, and C_S is discharged. During phase X_2, the opamp along with C_S and C_F operates as an inverting amplifier, with the bottom plate of the upper Cs is switched from analog ground (AGND) to V_{REFP}, while the lower Cs is switched from AGND to V_{REFN}. As a result, the residue V_{res} is subtracted by $(V_{REFP} - V_{REFN})$ if the previous bit decision DS2 was high. At the end of phase X_2, due to the charge sharing of C_R and C_S, the voltage across C_R becomes $(V_{REFP} - V_{REFN})/2$. Hence, as the operation continues, a sequence of the binary weights $1/2^N$ ($N = 1, 2, \ldots, M_2$) is generated by charge sharing for M_2 periods. The previous bit decision D_{S2} decides whether to subtract or add back the weight to the residue. Thus, a successive binary search is performed. After $M_1 + M_2$ clock periods, V_{RES} is bounded by $V_{REF}/2^{M_2}$. The effective 1 least-significant-bit (LSB) quantization error is thus $V_{RES}/2^{(M_2+M_1)}$. The two-phase nonoverlapping clocks Φ_1/Φ_2, S_1/S_2, and X_1/X_2 used to control the switched-capacitor circuits in each step are illustrated in Figure 4.9. They can be generated by simple logic circuits. No calibration was needed for the capacitance mismatch in the 2-C DAC. Simulations show no SQNR loss for a 0.1% capacitance mismatch and only a 3 dB loss with a 0.2% mismatch. Since the C_R sizes are large, 0.1% accuracy is achievable by careful layout.

4.2.3 Nonideal Effects

The design Eqs. (4.7)–(4.10) ignored all nonideal effects. All building blocks were assumed ideal. When the circuit is implemented, the major nonidealities have to be estimated and mitigated. For a single-loop IADC1, the counting conversion is based on a first-order modulation, which is known to be tolerant toward offset. The input-referred offset voltage of the comparator does not degrade the SQNR. The extended counting, however, is an algorithmic conversion, and a dc offset causes nonlinearity [18]. Large comparator offset voltage causes more SQNR degradation. Figure 4.13a shows the simulated SQNR versus comparator's input-referred offset voltage with an input amplitude of −3 dBFS. Simulations show that the SQNR remains above 108 dB when the offset voltage is not higher than 1 mV. In [18], trilevel internal quantizer is used in the loop to mitigate the offset effect. As in [17], auto-zeroing is also incorporated in the proposed scheme to lower the offset. The auto-zeroing preamplifier of the one-bit quantizer stores the input-referred offset during Φ_2. The stored offset voltage is then canceled during the comparison

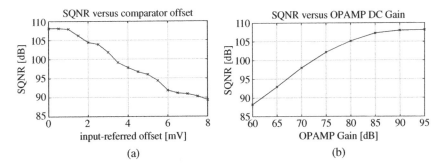

Figure 4.13 (a) Simulated SQNR versus IADC1 comparator's offset voltage when the input amplitude is −3 dBFS. (b) Simulated SQNR versus opamp DC gain when the input amplitude is −3 dBFS.

phase. It is easy to keep the thermal noise power contributed by the comparator negligible in comparison with the amplitude of the LSB.

MASH schemes are also sensitive to the opamp dc gain [1]. The opamp is required to have high dc gain to make the analog integrator ideal enough, so that the IADC1's quantization error E_1 becomes negligible after extended counting. The simulated SQNR versus the opamp dc gain is plotted in Figure 4.13b. In order to achieve an 18-bit SQNR, the required opamp gain is 85 dB. A two-stage opamp was chosen to implement the integrator and buffer in the 2-C SAR. Figure 4.14a shows the topology of the two-stage opamp circuit. Since the opamp sees very different capacitive loads during the two-step operation, Miller capacitor cascode compensation was adopted. It stabilizes the opamp by making the dominant pole independent of the capacitive loading. The simulation assumes a typical opamp performance, 86 dB DC gain, 20 MHz gain-bandwidth (GBW), 63-degree phase margin, and 16.8 µW power consumption.

Since the input sampling capacitor size is large, the mismatch of the integrator coefficient of the IADC1 does not cause degradation. The sampled kT/C thermal noise introduced by the 2-C SAR is negligible when referred to the input of the complete two-step ADC.

4.2.4 Measured Performance

The proposed hybrid ADC was implemented in a 0.18-µm CMOS technology. The total active area was only 0.27 mm². Figure 4.14b shows the die micrograph of the prototyped IADC. The ADC was tested with sinusoid signals, and Figure 4.15 shows the measured power spectral densities (PSDs) for input signals at 170 and 800 Hz. The PSDs after the first step, and after the two-step operations are compared. For the first step only (OSR = 256), the IADC1 achieves 52.1 dB SNR, which

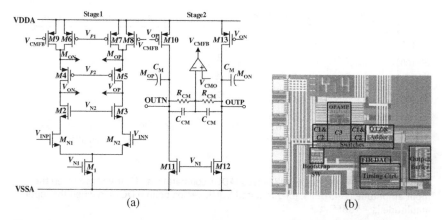

(a)   (b)

Figure 4.14 (a) Two-stage opamp circuit. Simulations show 86 dB dc gain, 20 MHz GBW product, 63-degree phase margin, and 16.8 μW power consumption. (b) Die photo.

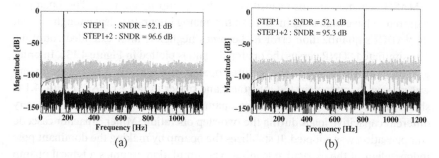

(a)   (b)

Figure 4.15 Measured spectra at signal frequencies: (a) 170 Hz and (b) 800 Hz.

is limited by the quantization noise of IADC1. The second-step extended counting (total OSR = 266) raises the total SNDR upto 96.6 dB. The overall SNR of both steps is limited by the sampled kT/C noise.

Operated with a 1.5 V supply, the differential full-scale voltage is 2 V_{PP}. Figure 4.16a plots the SNR and SNDR versus the input voltage amplitude. The measured dynamic range is 100.2 dB, and the peak SNR is 97.1 dB. The measured SNDR and SFDR are plotted versus the input frequency in Figure 4.16b.

Since the signal with a 1.2 kHz bandwidth (BW) suffers from flicker noise, using the chopper can reduce the noise significantly. Tested by shorting the ADC input terminals to the common-mode voltage of 0.75 V, the measured PSD shown in Figure 4.17a compares noise conditions with chopper switch on against off. Turning on the chopper stabilization, the measured output noise voltage is reduced from 8.34 to 4.12 μV_{RMS}, which is about a 6 dB improvement.

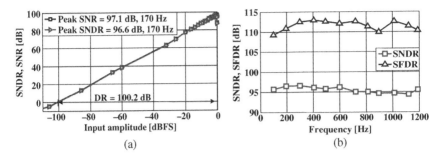

(a)   (b)

Figure 4.16 (a) Measured SNR and SNDR versus differential voltage amplitude. (b) Measured SNDR/SFDR versus input frequency.

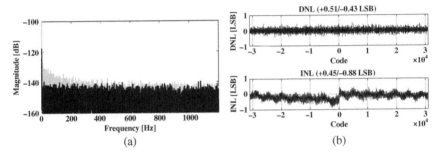

(a)   (b)

Figure 4.17 (a) Measured PSD when chopper was turned on or off. (b) Measured INL/DNL.

The measured differential nonlinearities (DNL) and integral nonlinearities (INL) are $+0.51/-0.43$ LSB and $+0.45/-0.88$ LSB, which are shown in Figure 4.17b. Operated at sampling frequency 642 kHz, the power dissipation is 33.2 µW, 76% of which is dissipated by the opamp of the integrator. The measured results are summarized in Table 4.1 and are compared with state-of-art hybrid IADCs and single-loop IADCs. The prototyped ADC achieves a Walden FoM_W of 0.25 pJ/conversion-step and a Schreier FoM_S of 175.8 dB. As these results show, the hardware reusing and 2-C SAR achieve a high power efficiency and very compact chip area.

In this section, the design of IADCs and the extended counting scheme to enhance the resolution of ADCs were reviewed and analyzed. Implementing a 2-C DAC, an IADC reconfigured as a 2-C SAR ADC was proposed to perform extended counting. Only one opamp was needed, reused in the two steps. The IADC was implemented in 0.18-µm CMOS, and it achieved 96.6 dB SNR, with a Schreier FoM of 175.8 dB and Walden FoM of 0.25 pJ/conversion-step. The prototype confirmed the usefulness of design concept, achieving great performance with excellent energy efficiency.

Table 4.1 Comparison with recent hybrid and single-loop IADCs.

Parameter	This work [12]	[23]	[24]	[25]	[26]	[27]	[10]	[17]	[28]	[19]	[20]	[21]	[22]	[6]
		TCAS1 2020	JSSC 2020	JSSC 2018	SSCL 2019	JSSC 2019	SSCL 2018	JSSC 2017	JSSC 2015	ISSCC 2016	ISSCC 2013	TCAS1 2012	TCAS1 2010	JSSC 2010
Architecture	IADC1 + Binary counting	Inter-leaved IADC2	IADC2 Cap scaling	Int. slice IADC2	CT IADC FIR DAC	IADC2 +Exp. Count	IADC1 +10-b cyclic	IADC1 +Multi-Slope	IADC2 +IADC1	Zoom ADC	Single IADC2	2-Step CT IADC	IADC2 +10-b cyclic	+11-b SAR
Process (nm)	180	180	180	180	180	65	180	180	65	160	160	180	180	180
Area (mm²)	0.27	0.45	0.66	0.363	0.175	0.13	0.72	0.32	0.2	0.16	0.45	0.337	0.5	3.5
V_{DD} (V)	1.5	3	1.8	3	3	1.2	3	1.5	1.2	1.8	1	1.2/1.8	2	1.8
Max. diff. Amp. (V_{PP})	2	4	2.94	3.8	3.19	2.14	6	2	2.2	3.5	0.7	0.7	3.6	2
Samp. Freq. (Hz)	642k	30M	400k	30M	32M	10.24M	55M	642k	192k	11M	750k	320k	115M	45.2M
Bandwidth (kHz)	1.2	100	2.04	100	100	20	625	1	0.25	20	0.667	4	11500	500
Power (µW)	33.2	1347	25.4	1098	1270	550	27700	34.6	10.7	1650	20	34.8	48000	38100
SNR_{MAX} (dB)	97.1	85.2	96	88.2	86	—	97.3	98.4	—	104.4	—	—	72	86.3
$SNDR_{MAX}$ (dB)	96.6	85.1	95.5	86.6	83	100.8	96.6	96.8	90.8	98.3	81.9	75.9	72	86.3
DR (dB)	100.2	87.2	102.2	91.5	91.5	101.8	100.1	99.7	99.8	107.5	81.9	85.5	73	90.1
$FoM_{S,DR}$ [a] (dB)	175.8	165.9	181.1	171.1	170.4	176.4	173.6	174.6	173.5	178.3	157.1	166.1	156.8	161.3
$FoM_{S,SNDR}$ [b] (dB)	172.2	163.8	174.4	166.2	161.9	175.4	170.1	171.7	164.5	169.1	157.1	156.5	155.8	157.5
FoM_W (pJ/conv.) [c]	0.25	0.46	0.13	0.31	0.55	0.15	0.40	0.31	0.76	0.61	1.47	0.85	0.64	2.26

a) $FoM_{S,DR} = DR + 10 \log_{10} (BW/Power)$.
b) $FoM_{S,SNDR} = SNDR + 10 \log_{10} (BW/Power)$.
c) $FoM_W = Power/(2^{(SNDR-1.76)/6.02} \cdot 2\,BW)$.

4.3 The Zoom Incremental ADC

4.3.1 Two-Stage 0–L IADCs

An alternative realization of a two-stage IADC is the 0–L MASH configuration. The basic block diagram of such a converter is shown in Figure 4.18. The coarse Nyquist-rate converter contributes a few N MSBs (typically, $N = 5$ to 6 bits) to the overall resolution. This can reduce the OSR of the next IADC stage. However, the overall accuracy is limited by that of the coarse ADC and DAC, and of the subtraction needed to generate the residue error for the fine IADC.

An ingenious modification, the *zoom incremental ADC* which mitigates these limitations, was proposed in Refs. [29–32]. Its block diagram is shown in Figure 4.19. Here, the conversion error of the coarse ADC is canceled by the IADC. The output voltage in the coarse ADC is also used to establish the reference voltages of the DAC used in the IADC in such a way that they track the input voltage V_{in} and reducing the IADCs quantization noise. Under ideal operations, the errors of the coarse ADC do not directly affect the overall accuracy. It is

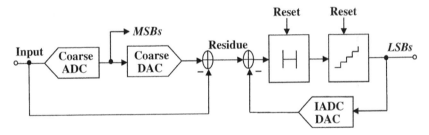

Figure 4.18 The block diagram of a 0–L incremental ADC.

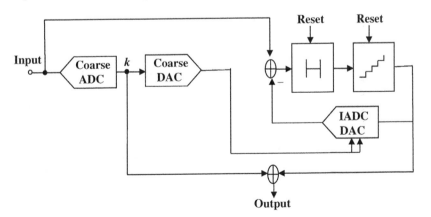

Figure 4.19 The block diagram of a 0–L zoom ADC.

determined only by the conversion accuracy of the IADC, which is greatly enhanced by the zoom operation. Specifically, let the coarse ADC have a resolution of N bits, and operate with reference voltages 0 and V_{ref}. Let its internal digital output be $Y_c = k$, indicating that its input signal V_{in} is between the references $k \cdot V_{LSB}$ and $(k+1) \cdot V_{LSB}$, where the LSB voltage is $V_{LSB} = 2^{-N} \cdot V_{ref}$. Then the coarse DAC output will be $k \cdot V_{LSB}$.

Assume next that the IADC operates with a single-bit quantizer and DAC, which is often the case in IADCs to achieve high linearity of operation. Then, the DAC of the IADC has only two reference voltages. Using $k \cdot V_{LSB}$ and $(k+1)V_{LSB}$ as these reference voltages, the swing of the IADC input signal is at most V_{LSB}. This allows high out-of-band gain and high accuracy for the IADC. The error of the coarse conversion is $V_{ec} = V_{in} - k \cdot V_{LSB}$. By subtracting the coarse DAC output from V_{in}, the coarse error residue $V_{ec} = V_{in} - k \cdot V_{LSB}$ can be entered into the IADC and converted with a high accuracy. For high-accuracy IADC with unity STF, the signal component of the IADC output Y_{IADC} will be the digital replica of this input V_{ec}. Hence, adding Y_c and Y_{IADC} cancels V_{ec} in the overall output. The error of the coarse conversion is thus replaced by the shaped and reduced quantization noise of the IADC.

The swing of the input signal to the IADC loop filter is thus ideally limited by the scheme to be less than V_{LSB}. This small swing reduces the input quantization noise by a factor close to 2^{-N}, enhancing the SQNR. The reduced input signal also improves the linearity of the loop filter and allows the use of simpler and less power-hungry amplifiers. The described scheme allows the input references of the fine IADC to *zoom* in on the vicinity of the current value of the input signal, giving the name of these ADCs.

Due to the nonidealities (offset and mismatch errors) of the coarse ADC, the input signal of the IADC may exceed the range specified by the reference voltages. This can be prevented by replacing the input range from $k \cdot V_{LSB}$ to $(k+1) \cdot V_{LSB}$ with an enlarged range of $(k-1) \cdot V_{LSB}$ to $(k+2) \cdot V_{LSB}$. The input range will now be $3V_{LSB}$ wide, which allows greatly relaxed requirements for the coarse ADC. However, the quantization noise reduction will now be only by a factor of 2^{-N+2}.

The loss of resolution due to the redundancy described can be mitigated by using a multi-bit quantizer in the IADC, combined with the original single-bit DAC. Figure 4.20 shows the resulting IADC DAC output signals, along with V_{in} and the coarse ADC output k, for 1-bit and 2-bit IADC quantizers [32].

In the first description of the zoom ADC [29], the operations of the coarse ADC and the IADC were sequential. The IADC has started the conversion only after the coarse ADC has found the range of the input voltage V_{in}. This limited the operation to the conversion of very slowly varying signals, such as those originating from temperature or capacitance measurements. Subsequently, it was found that for a limited signal BW, the two converters forming the zoom ADC may be operated simultaneously [30]. The criterion for the maximum signal

Figure 4.20 DAC output waveforms for zoom IADCs: (a) with single-bit quantizer and (b) with 2-bit quantizer. Source: Eland et al. [32]/IEEE.

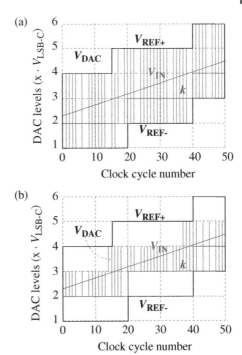

frequency is that the variation of the input voltage ΔV_{in} due to the delay between the operations of the two ADC stages should not exceed the stable input range of the IADC converter. The converter with the improved scheme was named *dynamic zoom ADC*. It enabled the design of an audio ADC with a 103 dB SNDR using a clock frequency of 11.29 MHz [30].

4.4 Zoom ADC Design Example

Reference [29] proposed the concept of the zoom IADC for the first time. In this work, the proposed zoom IADC is directly used as the fine converter of a two-step ADC, whereas, the residue was not computed by the coarse conversion and DAC, as shown in Figure 4.18. Rather, the results of the coarse conversion were used to dynamically adjust the references of the fine ADC such that the reference range tracks the current value of the input signals. Therefore, the residue computation no longer introduces any errors.

The proposed zoom IADC is shown in Figure 4.21b. The coarse ADC resolution was chosen to be 6 bits, and the fine IADC was designed to be a 2nd order IADC, which theoretically requires 300 cycles to achieve SQNR of 130 dB. The whole conversion is performed sequentially due to the fact the input is a static signal for instrumentation applications. During the coarse SAR conversion, the input is first

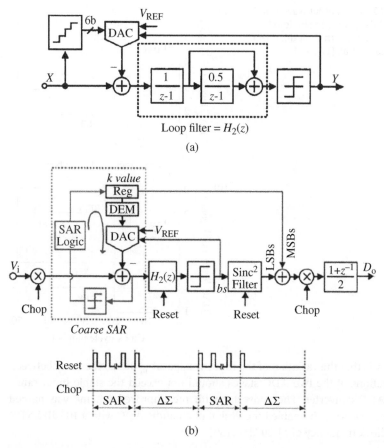

Figure 4.21 (a) Concept diagram of the zoom 2nd-order IADC with 1-bit quantizer. (b) Block diagram of the proposed zoom ADC, along with the timing. Source: Chae et al. [29]/IEEE.

sampled onto the capacitive DAC. After six SAR conversions, the coarse result k is found such that $A = kV_{\text{LSB, SAR}} < V_i < (k+1)V_{\text{LSB, SAR}}$. The digital result k is then stored in the SAR register and used to reconfigure the references used for the IADC. During the fine conversion, the input signal is digitized by the following 2nd-order IADC.

4.4.1 Nonideal Effects

The fine conversion range of the zoom IADC was made to be 2 LSBs of the coarse conversion, so that the offset and quantization error of the coarse ADC can be accommodated. This redundancy makes the input safely stay within the input range of the fine ADC and thus relaxed the required accuracy of the coarse ADC.

By combining the outputs of the two conversion steps, the digital output can be expressed as

$$Y_{OUT} = Y_c \cdot 2^{N-1} + \text{YIADC}$$

where, N is the bit resolution of fine IADC, and Y_c and Y_{IADC} are the outputs of the coarse and fine ADCs, respectively.

Since the output swing of the integrator used in the zoom IADC is greatly relaxed thanks to the 6-bit coarse ADC, the proposed zoom ADC requires an output swing of only 4.5% of the reference voltage. As a result, the OTA design has relaxed settling and slewing requirements. The settling behavior follows a single pole response and the introduced error results in a fixed gain error in the integrator.

The decimation filter was chosen to have the sinc^2 transfer function shown below:

$$H(z) = \frac{1}{D^2} \left(\frac{1 - z^{-D}}{1 - z^{-1}} \right)^2$$

The sinc^2 filter has notches at the integer multiples of the fs/D. The power line noise can, therefore, be suppressed by appropriately choosing the sampling frequency and decimation ratio. In this work, the sampling frequency $f_s = 25.6$ kHz and the decimation factor $D = 512$ with ADC cycles of $N = 1024$ were chosen, which created the first notch at the 50 Hz European power line frequency.

The element mismatches in the DAC impact the overall ADC's linearity. Dynamic element matching (DEM) was applied to mitigate the effect. The residue error E of the DEM after N cycles is bounded [32] by

$$E = \left| \frac{1}{N} \cdot \sum_{i=1}^{N} \delta_i \right| < \frac{1}{N} \cdot \sqrt{2^M} \cdot \delta_{max}$$

where δ_{max} is the worst-case mismatch. With the initial process capacitor (160 fF) having a mismatch of 0.05%, 6 bits DAC elements, and 1024 ADC clock cycles, ± 2 ppm can be achieved.

The sampling capacitor sizing is determined by the kT/C noise of the fine IADC, since there is no residue computation due to the zooming operation. In this work, the sampling capacitor of the fine IADC was chosen to be 10.2 pF which was then divided into 64 unit capacitors of 160 fF, with 1024 ADC cycles.

4.4.2 Measured Performance

The prototype consumes 3.5 µA under 1.2 to 2 V supply voltage, achieving INL of 6 ppm with DEM applied to the capacitive DAC. The output noise is 0.65 µV$_{RMS}$ with a conversion time of 40 ms, resulting in an SNR of 119.8 dB with ± 0.9 V differential input, as shown in Figure 4.22a,b. The ADC's power supply rejection ratio (PSRR) is about 120 dB at DC and 103 dB at 50 Hz. The Schreier FoM achieves 182.7 dB with 40 ms conversion time.

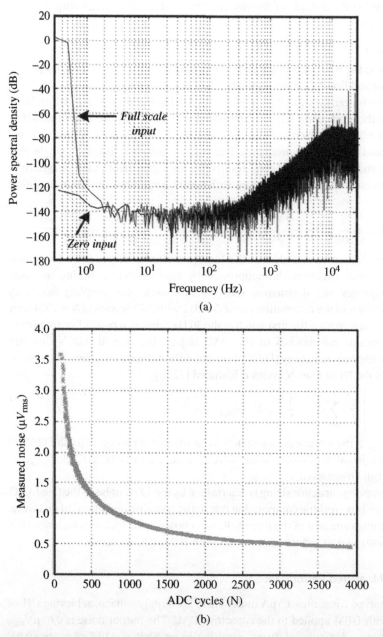

Figure 4.22 (a) Measured output spectrum of the zoom IADC with DEM on. (b) Measured noise referred to the input over the ADC cycles.

References

1 R. Schreier and G. C. Temes, *Understanding Delta-Sigma Data Converters.* Pascataway, NJ, USA: IEEE Press/Wiley, 2005.

2 I. Fujimori et al., "A 90-dB SNR 2.5-MHz output-rate ADC using cascaded multibit delta-sigma modulation at 8x oversampling ratio," *IEEE Journal of Solid-State Circuits*, vol. 35, no. 12, pp. 1820–1828, Dec. 2000.

3 T. C. Caldwell and D. A. Johns, "Incremental data converters at low oversampling ratios," *IEEE Transactions on Circuits and Systems I: Regular Papers*, vol. 57, no. 7, pp. 1525–1537, Jul. 2010.

4 J. Robert and P. Deval, "A second-order high-resolution incremental A/D converter with offset and charge injection compensation," *IEEE Journal of Solid-State Circuits*, vol. 23, no. 3, pp. 736–741, Jun. 1988.

5 R. Harjani and T. A. Lee, "FRC: a method for extending the resolution of Nyquist rate converters using oversampling," *IEEE Transactions on Circuits and Systems II: Express Briefs*, vol. 45, no. 4, pp. 482–494, Apr. 1998.

6 A. Agah et al., "A high-resolution low-power oversampling ADC with extended-range for bio-sensor arrays," *IEEE Journal of Solid-State Circuits*, vol. 45, no. 6, pp. 1099–1110, Jun. 2010.

7 J.-H. Kim et al., "A 14b extended counting ADC implemented in a 24MPixel APS-C CMOS image sensor," *IEEE International Solid-State Circuits Conference (ISSCC)*, pp. 390–392, Feb. 2012.

8 P. Rombouts, W. De Wilde, and L. Weyten, "A 13.5-b 1.2-V micropower extended counting A/D converter," *IEEE Journal of Solid-State Circuits*, vol. 36, no. 2, pp. 176–183, Feb. 2001.

9 P. Rombouts, P. Woestyn, M. De Bock, and J. Raman, "A very compact 1 MS/s Nyquist-rate A/D converter with 12 effective bits," *IEEE European Solid-State Circuits Conference (ESSCIRC)*, 2012.

10 T. Katayama, S. Miyashita, K. Sobue, and K. Hamashita, "A 1.25 MS/s two-step incremental ADC with 100-dB DR and 110-dB SFDR," in *IEEE Solid-State Circuits Letters*, vol. 1, no. 11, pp. 207–210, Nov. 2018.

11 Y. Zhang, C.-H. Chen, T. He, K. Sobue, K. Hamashita, and G.C. Temes, "A two-capacitor SAR-assisted multi-step incremental ADC with a single amplifier achieving 96.6 dB SNDR over 1.2 kHz BW," *IEEE Custom Integrated Circuits Conference (CICC)*, San Jose, CA, USA, 2017.

12 S.-C. Kuo, J.-S. Huang, Y.-C. Huang, C.-W. Kao, C.-W. Hsu, and C.-H. Chen, "A multi-step incremental analog-to-digital converter with a single-opamp and two-capacitor SAR extended counting," *IEEE Transactions on Circuits and Systems I: Regular Papers*, vol. 68, no. 7, pp. 2890–2899, Jul. 2021.

13 R. Suarez, P. R. Gray, and D. A. Hodges, "All MOS charge redistribution analog-to-digital conversion techniques—part II," *IEEE Journal of Solid-State Circuits*, vol. 10, no. 6, pp. 379–384, Dec. 1975.

14 C.-H Chen, Y. Zhang, J. Ceballos, and G. C. Temes, "Noise-shaping SAR ADC using three capacitor," *Electronics Letters*, vol. 49, no. 3, pp. 182-184, 2013.

15 Y. Zhang, C.-H Chen, T. He, and G. C. Temes, "Multi-step incremental ADC with extended binary counting," *Electronics Letters*, vol. 52, no. 9, pp. 697–699, 2016.

16 O. Oliaei, "Sigma-delta modulator with spectrally shaped feedback," *IEEE Transactions on Circuits and Systems II: Express Briefs*, vol. 50, no. 9, pp. 518–530, Sept. 2003.

17 Y. Zhang, C.-H Chen, T. He, and G. C. Temes, "A 16b multi-step incremental analog-to-digital converter with single-opamp multi-slope extended counting," *IEEE Journal of Solid-State Circuits*, vol. 52, no. 4, pp. 1066–1076, Apr. 2017.

18 B. Ginetti, P. G. A. Jesper, and A. Vandemeulebroecke, "A CMOS 13-b cycle RSD A/D converter," *IEEE Journal of Solid-State Circuits*, vol. 27, no. 7, pp. 957–965, Jul. 1992.

19 B. Goenen, F. Sebastiano, R. Veldhoven, and K. Makinwa, "A 1.65 mW 0.16 mm² dynamic zoom-ADC with 107.5 dB DR in 20 kHz BW," *IEEE ISSCC Digest of Technical Papers*, pp. 282–283, Feb. 2016.

20 C. Chen, Z. Tan, and M. A. P. Pertijs, "A 1V 14b self-timed zero-crossing-based incremental ΔΣ ADC," *IEEE ISSCC Digest of Technical Papers*, pp. 274–275, Feb. 2013.

21 S. Tao and A. Rusu, "A power-efficient continuous-time incremental sigma-delta ADC for neural recording systems," *IEEE Transactions on Circuits and Systems I: Regular Papers*, vol. 62, no. 6, pp. 1489–1498, Jun. 2015.

22 C. C. Lee and M. P. Flynn, "A 14b 23 MS/s 48 mW resetting ΔΣ ADC," *IEEE Transactions on Circuits and Systems I: Regular Papers*, vol. 58, no. 6, pp. 1167–1177, Jun. 2011.

23 P. Vogelmann, J. Wagner, and M. Ortmanns, "A 14b twofold time-interleaved incremental ΔΣ ADC using hardware sharing," *IEEE Transactions on Circuits and Systems I: Regular Papers*, vol. 67, no. 11, pp. 3681–3692, Nov. 2020.

24 S. Mohamad, J. Yuan, and A. Bermak, "A 102.2-dB, 181.1-dB FoM extended counting analog-to-digital converter with capacitor scaling," *IEEE Journal of Solid-State Circuits*, vol. 55, no. 5, pp. 1351–360, May 2020.

25 P. Vogelmann, J. Wagner, M. Haas, and M. Ortmanns, "A dynamic power reduction technique for incremental ΔΣ modulators," *IEEE Journal of Solid-State Circuits*, vol. 54, no. 5, pp. 1455–1467, May 2019.

26 M. A. Mokhtar, P. Vogelmann, M. Haas, and M. Ortmanns, "A 94.3-dB SFDR, 91.5-dB DR, and 200-kS/s CT incremental delta–sigma modulator with differentially reset FIR feedback," *IEEE Solid-State Circuits Letters*, vol. 2, no. 9, pp. 87–90, Sep. 2019.

27 B. Wang, S. Sin, S. U. F. Maloberti, and R. P. Martins, "A 550-μW 20-kHz BW 100.8-dB SNDR linear-exponential multi-bit incremental ADC with 256 clock cycles in 65-nm CMOS," *IEEE Journal of Solid-State Circuits*, vol. 54, no. 4, pp. 1161–1172, Apr. 2019.

28 C.-H. Chen, Y. Zhang, T. He, P. Chiang, and G. C. Temes, "A Micro-power two-step incremental analog-to-digital converter," *IEEE Journal of Solid-State Circuits*, vol. 50, no. 8, pp. 1796–1808, Aug. 2015.

29 Y. Chae, K. Souri, and K. A. A. Makinwa, "A 6.3 μW 20 bit incremental zoom-ADC with 6 ppm INL and 1 μV offset," *IEEE Journal of Solid-State Circuits*, vol. 48, no. 12, pp. 3019–3027, Dec. 2013.

30 B. Gönen, F. Sebastiano, R. Quan, R. van Veldhoven, and K. A. A. Makinwa, "A dynamic zoom ADC with 109-dB DR for audio applications," *IEEE Journal of Solid-State Circuits*, vol. 52, no. 6, pp. 1542–1550, Jun. 2017.

31 S. Karmakar, B. Gönen, F. Sebastiano, R. van Veldhoven, and K. A. A. Makinwa, "A 280 μW dynamic-zoom ADC with 120 dB DR and 118 dB SNDR in 1 kHz BW," *IEEE Journal of Solid-State Circuits*, vol. 53, no. 12, pp. 3497–3507, Dec. 2018.

32 E. Eland, S. Karmakar, B. Gönen, R. van Veldhoven, and K. A. A. Makinwa, "A 440-μW, 109.8-dB DR, 106.5-dB SNDR discrete-time zoom ADC With a 20-kHz BW," *IEEE Journal of Solid-State Circuits*, vol. 56, no. 4, Apr. 2021, pp. 207–1215.

5

Design Examples

Design examples are illustrated in this chapter, from top-level system design down to switched-capacitor (SC) circuit implementation. The measured performance and figure-of-merit (FoM) of each design example are compared with recent state-of-the-art works. The first example demonstrates a third-order one-bit single-loop incremental analog-to-digital converter (IADC) [1]. The second example describes a first-order IADC with coarse and fine multi-slope extended counting, which achieves 96 dB with excellent FoM [2]. The third example is a two-step IADC, which reuses the hardware of a second-order IADC to achieve third-order noise-shaping performance [3]. The third example is a two-step continuous-time (CT) IADC and signal-to-noise ratio (SAR) two-step analog-to-digital converter (ADC) [4]. The last example illustrates a hardware reusing multistage noise shaping (MASH) 2-1 IADC, which achieves fifth-order noise-shaping performance [5].

5.1   A Third-Order 22-Bit IADC

As the first example, the design of a single-stage IADC, a third-order single-bit data converter [1] will be discussed. The block diagram of the noise-shaping loop is shown in Figure 5.1; it uses a cascade-of-integrators (CoI) feedforward loop filter with an extra feedforward path from the input to the quantizer. This added path ensures that the input signal does not pass through integrator, and thus reduces the harmonic distortion.

The SC circuit implementing it is illustrated in Figure 5.2. In order to avoid dynamic as well as static nonideal effects introduced by a multi-bit digital-to-analog converter (DAC), single-bit quantization was used. The scaling coefficients chosen were $a = [1.4\ 0.99\ 0.47]$, $b = 0.5674$, and $c = [0.5126\ 0.3171]$.

Figure 5.3 plots the mean square power spectral density (PSD) of the quantization noise as a function of the dc input signal u. Here, u normalized to V_{ref}.

Incremental Data Converters for Sensor Interfaces, First Edition. Chia-Hung Chen and Gabor C. Temes.
© 2024 The Institute of Electrical and Electronics Engineers, Inc. Published 2024 by John Wiley & Sons, Inc.

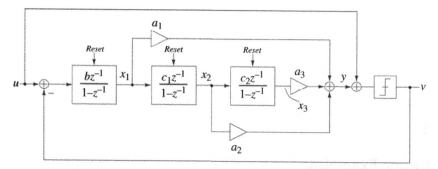

Figure 5.1 Block diagram of the 22-bit third-order IADC. Source: Adapted from Quiquempoix et al. [1].

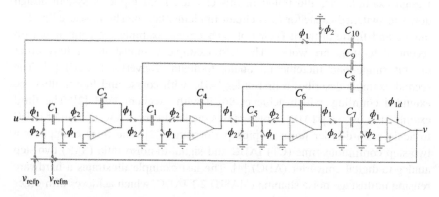

Figure 5.2 Single-ended switched-capacitor schematic. Source: Adapted from Quiquempoix et al. [1].

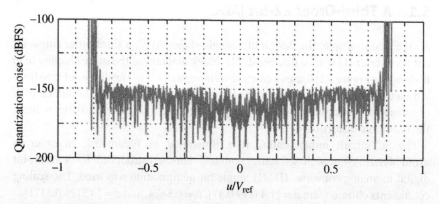

Figure 5.3 Quantization noise power as a function of u/V_{ref} for $M = 1024$.

The oversampling ratio was $M = 1024$. When $|u|$ approaches V_{ref}, the quantizer will overload, and the noise becomes large. This is true for all IADCs, as well as delta-sigma ADCs. Note that there are no other spurs indicating idle tones, since the reset operations prevent the occurrence of periodic signals with long periods, and the digital filter suppresses high-frequency tones.

To avoid the overload caused by large input signals close to $|u| = V_{ref}$, an attenuator with a gain of 2/3 was included in the input stage. To make this gain factor accurate even if there are element value mismatches, a dynamic element matching scheme was implemented. The circuit realizing this scheme is shown in Figure 5.4. In one clock period, all six switched input capacitors deliver a charge proportional to the DAC output V_{dac}, but only four of them deliver the $C_1 \cdot u$ charge. This is equivalent to the desired scale factor of 2/3 for u. The roles played by each of the capacitors are rotated from clock period to clock period. Thus, the noise introduced by the mismatch errors of the capacitors is converted into an out-of-band periodic noise.

To cancel the dc offset, an enhanced form of chopping, named *fractal sequencing*, was implemented. It is illustrated in Figure 5.5. The reason is that simple chopping is inadequate in the CoI circuit used here. To verify this statement,

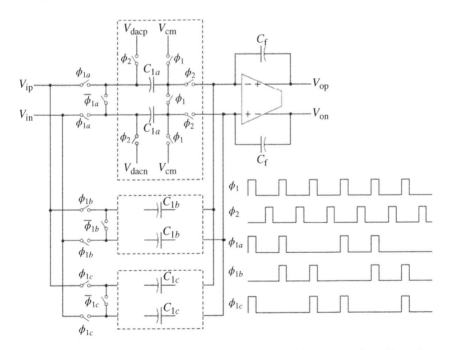

Figure 5.4 The rotating capacitor input circuit. The dotted frames contain replicas of the circuit containing C_{1a}.

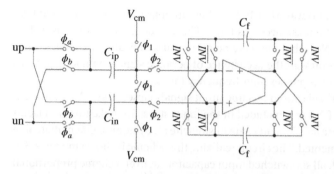

Figure 5.5 The input integrator with offset compensation using fractal sequencing.

assume that a 1-mV offset exists at the input of the first integrator (INT1) and that unity gain factors were used for all three integrator stages. Then the output sequence of the INT1 (in mV) will be {1, −1, 1, −1, ...}; the second stage output is {1, 0, 1, 0, ...}; and the third one {1, 1, 2, 2, 3, 3, ...}—diverging with time. In fractal sequencing, the control signal INV ensures that the input signal is always integrated with the same sign, while the input offset is toggled in such a way that after M oversampled cycles the offset at the output of the last integrator is canceled.

The sequence of simple chopping {+ − + − ...} can be described by the operator $S_1 = (+-)$. Here, a + denotes no inversion of the signal, while—denotes inversion, and the parentheses indicate that this pattern repeats indefinitely. By contrast, the fractal sequences S_k are based on the recursion relation $S_{k+1} = [S_k, -S_k]$. Thus, from the simple chopping sequence $S_1 = (+ -)$, higher-order sequences can be generated as shown in Eq. (12.17) below. The desired sequence for the IADC is S_L, where L is the number of the cascaded integrators in the loop. In the device discussed here, $L = 3$.

$$S_1 = (+-),$$
$$S_2 = [S_1, -S_1] = (+ - -+),$$
$$\vdots$$
$$S_{k+1} = [S_k, -S_k] \tag{5.1}$$

Figure 5.5 shows the input integrator with fractal sequencing. The switches INV and $\overline{\text{INV}}$ are operated by the S_3 fractal sequence. To maintain a consistent integration polarity, $\phi_a = \phi_1$ and $\phi_b = \phi_2$ when INV is low, but $\phi_a = \phi_2$ and $\phi_b = \phi_1$ when INV is high. Note that the chopping frequency used in the fractal sequence need not be the main clock frequency; it may be a subharmonic of the IADC clock. Figure 5.6 shows the normalized integrator output voltages

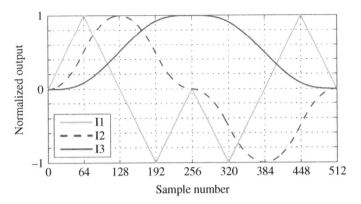

Figure 5.6 Integrator output voltages after fractal sequencing.

after fractal sequencing using $f_{chop} = fs/64$. Note that the dc output of the last integrator returns to 0 after $M/2 = 512$ clock periods.

The digital filter used to process the IADC output applied a modification of the sinc transfer function

$$H(z) = \prod_{i=1}^{4} \frac{1 - z^{-M_i}}{M_i(1 - z^{-1})} \tag{5.2}$$

where $M_i = \{512, 512, 512 - 2^6, 512 + 2^6\}$, which provided wide notches around the line frequency (Figure 5.7). These notches are useful for suppressing line-frequency noise even if the clock frequency or line frequency varies slightly.

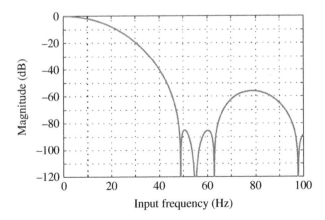

Figure 5.7 The gain response of the decimation filter.

5.2 A 16-Bit Multistep IADC with Single-Opamp Multi-Slope Extended Counting

In the extended counting [6] schemes described before, the extra order of noise shaping is obtained by recycling the residue voltages from the prior step. The accuracy is enhanced by delivering the residue error to the next stage. Hence, the fine quantization step is vulnerable to nonideal effects in switched capacitor circuits, such as switches charge injection, clock feed-through and coupling through parasitics in this transmission. Therefore, this step requires careful shielding and buffering of the quantization residue in order to achieve high accuracy.

Figure 5.8b shows a first-order IADC (IADC1) with a wideband topology, which operates like a dual-slope ADC with integration and subtraction intermixed in time. An IADC1 has the simplest form and is unconditionally stable within the non-overloading input range. Furthermore, it has a minimum thermal noise penalty factor of 1, compared to IADCs with high-order loops.

A multistep IADC will next be described, which extends the accuracy by using multi-slope extended counting [2]. It cancels the residue voltage of the coarse quantization without transmitting it to the next stage, and is hence, more robust than other extended counting ADCs.

The proposed scheme keeps the integrator in the same operating environment and cancels the residue error only by detecting the polarity of the quantization residue, rather than transferring the coarse residue error for finer quantization.

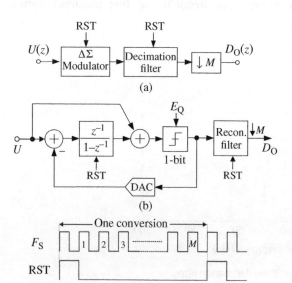

Figure 5.8 Block diagram of (a) a conceptual IADC and (b) a first-order IADC.

The scheme of the multi-slope extended counting technique is shown in Figure 5.9a, which demonstrates how the IADC is reconfigured in multistep operation to extend its accuracy. The timing diagram of the multistep operation is given in Figure 5.9b. Each conversion cycle between adjacent resets is split into three steps. The numbers of the clock periods assigned to the three steps are M_1, M_2, and M_3, respectively. During the first step, as shown in Figure 5.9a, the IADC uses a first-order $\Delta\Sigma$ loop with input feedforward architecture. For high linearity, a single-bit quantizer is used, and a finite-impulse-response (FIR) feedback path $F(z) = 0.5 + 0.5z^{-1}$, together with the compensation path $C(z) = 0.5z^{-1}$ is introduced to reduce the voltage step at the integrator output V_{RES}. A feedforward path $J(z) = 0.5z^{-1}$ is used to cancel the input signal content introduced by $C(z)$, and therefore, to maintain the signal transfer function (STF) of the IADC to be close to unity within the signal bandwidth (BW). As in a conventional feedforward IADC1, the integrator in the loop filter processes only the first-order shaped quantization noise, and it stores the quantization residue at its output V_{RES} after clock period M_1. Its magnitude is

$$|V_{RES}[M_1]| = \left| \sum_{i=1}^{M_1-1} V_{IN}[i] - F(z) \cdot V_{REF} \sum_{i=1}^{M_1-1} D_1[i] \right| \leq V_{REF} \tag{5.3}$$

Note that the quantization residue is bounded by the reference voltage V_{REF}. Next, $V_{RES}[M_1]$ is held at the integrator output for the second-step quantization. In the second step, as shown in Figure 5.9b, the input path is disconnected and the IADC is reconfigured to cancel the quantization residue by counting with a slope coefficient $G_2 = 1/M_2$. Accordingly, the DAC step size is scaled to V_{REF}/M_2. The residue $V_{RES}[M_1]$ at the integrator output is therefore re-quantized with a predetermined step size V_{REF}/M_2. At the end of the second step, the quantization residue is bounded by V_{REF}/M_2:

$$|V_{RES}[M_1 + M_2]| = \left| V_{RES}[M_1] - G_2 \cdot V_{REF} \sum_{i=1}^{M_2-1} D_2[i] \right| \leq \frac{V_{REF}}{M_2} \tag{5.4}$$

Similarly, for the final step, the circuit remains configured as a counting ADC, and it quantizes the quantization residue with a finer step size $V_{REF}/(M_2 \cdot M_3)$, resulting in a slope coefficient $G_3 = 1/(M_2 \cdot M_3)$, as shown in Figure 5.10c. The residue V_{RES} is then reduced to less than one LSB $= V_{REF}/(M_2 \cdot M_3)$ after M_3 additional steps:

$$|V_{RES}[M_1 + M_2 + M_3]| = \left| V_{RES}[M_1 + M_2] - G_3 \cdot V_{REF} \sum_{i=1}^{M_3-1} D_3[i] \right| \leq \frac{V_{REF}}{M_2 \times M_3} \tag{5.5}$$

Thus, in this process, the quantization residue is not recycled but is gradually canceled by a coarse and fine DAC step size. It is only connected to the quantizer for polarity decisions. This makes the process robust. From Eqs. (5.3)–(5.5), assuming that V_{IN} is constant during one conversion cycle, the digital representation of the average input signal $\overline{V_{IN}}$ can be estimated from

$$\overline{V_{IN}} = \frac{V_{REF}}{M_1 - 1}\left[F(z) \cdot \sum_{i=1}^{M_1-1} D_1[i] + G_2 \cdot \sum_{i=1}^{M_2-1} D_2[i] + G_3 \cdot \sum_{i=1}^{M_3-1} D_3[i]\right]$$
$$+ \frac{V_{REF}}{(M_1 - 1) \times M_2 \times M_3} \tag{5.6}$$

The first term on the right-hand side (RHS) of Eq. (5.6) determines the structure of the digital decimation filtering needed to reconstruct the input signal from the bit-streams $D_k[i]$ of the three steps. As Figure 5.9a shows, a single digital counter can be shared among all three steps for signal reconstruction. The second term on the RHS of Eq. (5.6) gives the equivalent quantization

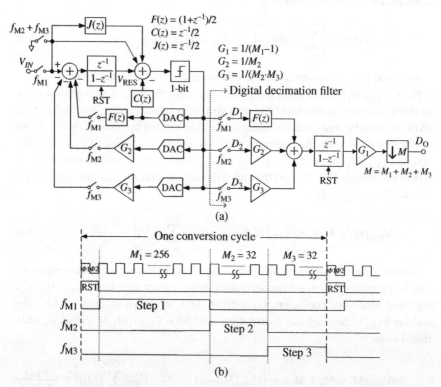

(a)

(b)

Figure 5.9 (a) Block diagram of the proposed multistep IADC with multi-slope extended counting. (b) Simplified timing diagram.

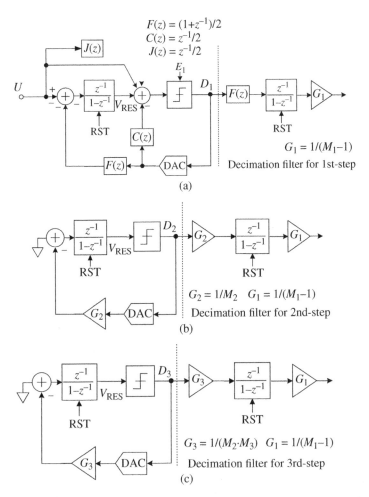

Figure 5.10 (a) First step: a first-order IADC with FIR feedback path. (b) Second step: circuit is reconfigured as a slope ADC with a coarse slope coefficient $G_2 = 1/M_2$. (c) Third step: circuit is reconfigured as a slope ADC with a reduced slope coefficient $G_3 = 1/(M_2 \cdot M_3)$.

error after three-step operation. It gives signal-to-quantization-noise ratio ($SQNR$) $\approx 20 \cdot \log_{10}(M_1 \cdot M_2 \cdot M_3)$. For such accuracy, a single-step IADC1 requires ($M_1 \cdot M_2 \cdot M_3$) clock periods for each conversion cycle, whereas, the proposed IADC needs only ($M_1 + M_2 + M_3$) ones. The effective number of bits (ENOB) of the SQNR for the example ADC was designed to be 18 bits. Since the practically achievable element matching accuracy is usually as high as 10 bits, the second and third steps were designed to realize 10-bit accuracy. As a result, $M_2 = M_3 = 2^5$

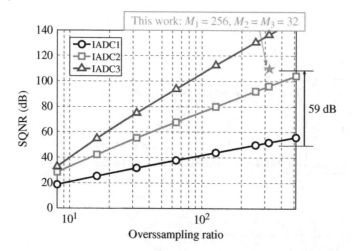

Figure 5.11 Comparison of the proposed architecture with single-step IADCs. IADC1: first-order IADC; IADC2: second-order IADC; and IADC3: third-order IADC.

gives the minimal total clock cycles for the additional two steps and M_1 is therefore assigned 2^8. This reduces the length of the conversion cycle to 320 ADC clocks, whereas, an IADC requires 2^{18} clock cycles to achieve the same SQNR.

A comparison of SQNRs as functions of the oversampling ratio (OSR) for the proposed architecture and single-step IADCs of different orders is illustrated in Figure 5.11. For the same OSR, the proposed multistep IADC using one single integrator can achieve higher SQNR than a single-step second-order IADC. This indicates a significant power efficiency enhancement.

The waveforms of the quantization residues of the steps at the integrator output are illustrated in Figure 5.12a, where the first part of the waveform for the first step was left out. The residue error from the first step is quantized in the second and third steps, until it becomes smaller than one least-significant-bit (LSB) due to the multi-slope extended counting. The FIR filtering feedback path $F(z)$ reduces the voltage step size at the integrator because it distributes the energy of the DAC signal over time. This relaxes the slewing requirements of the opamp used in the integrator, as illustrated in the histograms of the voltage step sizes shown in Figure 5.12b.

The prototype test chip was fabricated in a 0.18-μm CMOS process. It occupies an active area of 0.5 mm². The chip micrograph is shown in Figure 5.13. The die was packaged in a 48-pin QFN and clocked at 642 kHz. The proposed IADC consumes 35 μW. Both analog and digital blocks use a 1.5 V supply. A common-mode voltage of 0.75 V was used, and the differential full-scale range of the IADC is 2 V_{PP}. The achieved PSRR at 50 Hz line frequency is at least 102 dB.

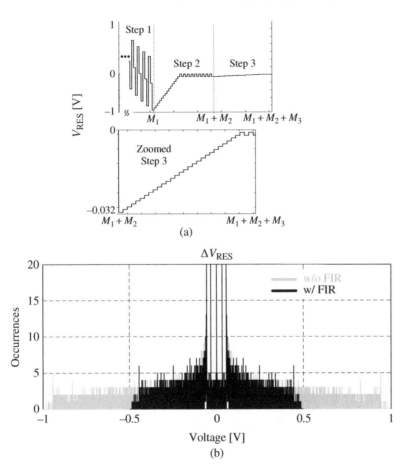

Figure 5.12 (a) Simulated waveform of the residue at the integrator output during each step. (b) Simulated histogram of the voltage step at the integrator output with and without FIR feedback.

Figure 5.14a shows the measured output PSDs for the three steps with a 170 Hz, −0.44 dBFS input sinusoid signal. It indicates that the IADC achieves a signal-to-noise-and-distortion-ratio (SNDR) of 52.2 dB over a 1 kHz BW during the first step. The extended multi-slope counting in the second and third steps then enhances the SNDR to 79.8 and 96.8 dB, respectively.

The signal-to-noise ratio (SNR) and SNDR are plotted as functions of the input amplitudes for 170 and 800 Hz inputs in Figure 5.15. The IADC achieves the dynamic range (DR) of 99.7 dB, and the peak SNR is 98.4 dB.

With the input, terminals shorted to the common-mode voltage of 0.75 V, the measured PSDs with the chopper turned on and off are illustrated in Figure 5.16.

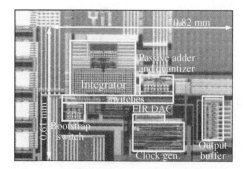

Figure 5.13 Chip micrograph of the prototype IADC.

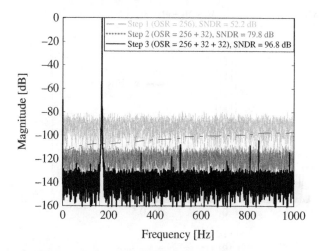

Figure 5.14 Measured PSDs for the three steps with a −0.44 dBFS sine wave input signal at 170 Hz frequency. Top: first step; middle: second step; and bottom: third step.

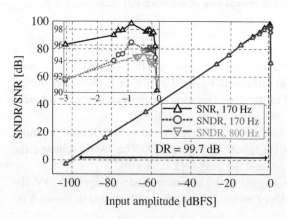

Figure 5.15 Measured SNR/SNDR versus the input amplitude for 170 and 800 Hz inputs.

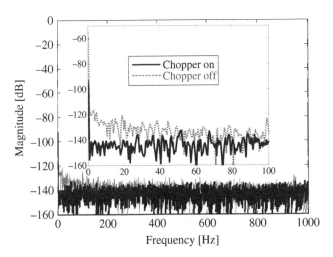

Figure 5.16 Measured spectra with the chopper turned on versus chopper turned off. Input terminals are shorted when measured.

The chopper stabilization reduced the flicker noise significantly and decreased the measured output noise from 7.54 to $3.76\,\mu V_{RMS}$. The measured dc offset was reduced from $-56\,dBFS$ (3.2 mV) to $-92\,dBFS$ (50.2 μV) by the chopper stabilization.

A dc input was swept over the full-scale range of the IADC, and the output noise was measured and plotted in terms of a 16-bit LSB, as shown in Figure 5.17. The spurs at two ends are due to the overloading of the IADC. The idle tones, which

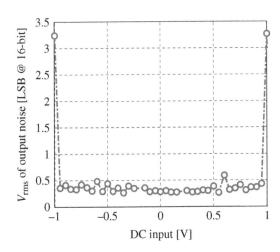

Figure 5.17 Measured output noise of the prototype IADC with dc input.

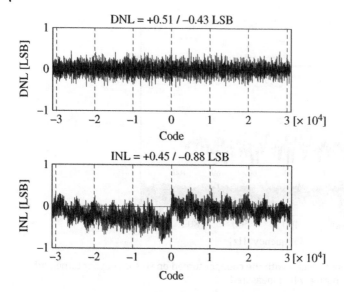

Figure 5.18 Measured INL/DNL in 16-bit accuracy LSBs.

commonly exist in low-order $\Delta\Sigma$ ADCs, are not found in the IADC. This can be intuitively understood as follows: when a dc input is applied to the IADC, the output sample taken from each reset window tends to stay the same except for the white thermal noise. Hence, no repetitive pattern forms, as does in $\Delta\Sigma$ ADCs. The differential nonlinearity (DNL) and integral nonlinearity (INL) errors determined by histogram testing with a sinusoidal-input signal are shown in Figure 5.18.

5.3 Multistep IADCs

5.3.1 Two-Step IADCs

As shown in Figure 3.5, we can improve the SQNR of an IADC by using a higher OSR or a higher-order modulator. For example, for an IADC2 with OSR = 64, doubling the OSR improves the SQNR by 15 dB, while increasing the order by 1 enhances the SQNR by 27 dB. Thus, it is more effective to increase the order of modulation than to raise the OSR. Unfortunately, increasing the order requires extra opamps. Besides, a higher-resolution internal quantizer is usually needed to make a higher-order modulator stable, and the complexity of the peripheral circuitry also increases. The power required increases accordingly, and the ADC becomes less efficient.

Next, a two-step architecture [3] will be described which avoids the excess power dissipation for high-resolution data conversion. Figure 5.19 shows the

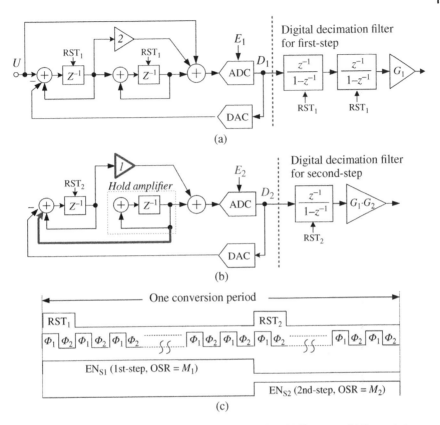

Figure 5.19 The proposed IADC2 in two-step operation. (a) First step. (b) Second step. (c) The simplified timing diagram.

z-domain model of the proposed two-step IADC2. During the first step, lasts for M_1 clock periods, the circuit is operated as a conventional IADC2 (Figure 5.19a). The residue voltage V_{RES} stored in the second integrator (INT2) after clock period M_1 is given by

$$V_{RES} = W_2[M_1] = \sum_{K=1}^{M_1-1} \sum_{i=1}^{K-1} U[i] - \sum_{K=1}^{M_1-1} \sum_{i=1}^{K-1} D_1[i] \qquad (5.7)$$

The direct-input feedforward modulator generates the residue voltage at the end of first conversion step for fine quantization.

To perform the second step (fine quantization), the analog modulator and the digital filter are reconfigured, as shown in Figure 5.19b. The INT2 now stops sampling and acts as a hold amplifier that feeds the residue voltage V_{RES} into the L-level quantizer and the INT1 is reset again and then samples the residue voltage

V_{RES} from INT2. The reconfigured circuits act as an IADC1 for the remaining M_2 clock periods. Analysis gives

$$\sum_{i=1}^{M_2-1} V_{RES} + E_2 = \sum_{i=1}^{M_2-1} D_2[i] \tag{5.8}$$

Since V_{RES} remains constant during the second step, it can be represented as

$$\sum_{i=1}^{M_2-1} V_{RES} = (M_2 - 1) \cdot V_{RES} \tag{5.9}$$

The quantization error E_1 of the first step IADC2 is canceled as in a MASH $\Delta\Sigma$ ADC, and only the final error E_2 remains after the two-step operation. While the input voltage U is sampled only during the first step, the average of the input signal $\tilde{U}$ can be redefined with M replaced by M_1

$$\tilde{U} \cong \frac{2}{M_1(M_1+1)} \sum_{j=1}^{M_1} \sum_{i=1}^{j} U[i] \tag{5.10}$$

After the two steps of conversion, the signals satisfy

$$\tilde{U} + \frac{2}{(M_1-1)(M_1-2)(M_2-1)} E_2 = \frac{2}{(M_1-1)(M_1-2)}$$
$$\times \left(\sum_{j=1}^{M_1-1} \sum_{i=1}^{j-1} D_1[i] + \frac{1}{(M_2-1)} \sum_{i=1}^{M_2-1} D_2[i] \right) \tag{5.11}$$

The decimation filter needed to reconstruct the bit streams of each step can be designed from the RHS of Eq. (5.11). For the first step, the decimation filter can be realized by two cascaded counters. For the second step, one of the counters can be reused. Thus, the IADC2's analog and digital hardware can be used in both steps with a simple reconfiguration. The equivalent quantization error of the two-step conversion can be estimated from

$$E_{21} = \frac{2}{(M_1-1)(M_1-2)(M_2-1)} \frac{V_{FS}}{L-1} \tag{5.12}$$

The SQNR at full-scale input amplitude is given by

$$\text{SQNR}_{21} = 20\log(V_{FS}/E_1) \approx 2 \cdot 20\log(M_1) + 20\log(M_2) + 20\log(L-1) - 6 \tag{5.13}$$

With a total OSR $= M = M_1 + M_2$, the optimal selection of the OSR values M_1 and M_2 in a two-step IADC is easily found. Defining the ratio $k = M_1/M_2$, we obtain $M_1 = kM/(k+1) M_1 = kM/(k+1)$ and $M_2 = M/(k+1) M_2 = M/(k+1)$. The quantization error of the two-step IADC can then be written in the form

$$E_{21} \approx \frac{2}{M_1^2 \cdot M_2} \frac{V_{FS}}{L-1} = \frac{2}{\frac{k^2 \cdot M^3}{(k+1)^3}} \frac{V_{FS}}{L-1} \tag{5.14}$$

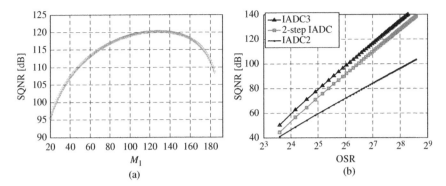

Figure 5.20 (a) Simulated SQNR versus OSR of the first step (M_1). The input amplitude is −6 dBFS. (b) Simulated SQNR versus OSR for a single-loop IADC3, IADC2, and the proposed two-step IADC2. All the IADCs are assumed to have a five-level quantizer and are tested at −6 dBFS input signal amplitude.

The minimum of the quantization error results by setting $k = 2$. The optimum OSRs of the two steps are then $M_1 = 2M_2 = 2M/3$. The maximum SQNR$_{21}$ is

$$\text{SQNR}_{21,\text{OPT}} \approx 3 \cdot 20 \log(M) + 20 \log(L - 1) - 20 \log(6) - 20 \log(9/4) \quad (5.15)$$

For a total OSR of 192, the SQNR of the two-step IADC versus M_1 is plotted in Figure 5.20a, which verifies that the maximum SQNR for $M_1 = 2M_2 M_1 = 2M_2 = 128$. It can be seen that the optimal ratio is not very sensitive to the exact value of k.

From Eq. (3.20), the SQNR of an IADC3 with the same conversion time is

$$\text{SQNR}_3 \approx 3 \cdot 20 \log(M) + 20 \log(L - 1) - 20 \log(6) \quad (5.16)$$

Comparison of SQNR$_{21}$ and SQNR$_3$ shows that the two-step IADC2's SQNR is 7 dB lower than that of an IADC3. However, its noise shaping is nearly one order higher than that of a single-loop IADC2. Figure 5.20b compares the simulated SQNR versus OSR curves for a single-loop IADC2, a single-loop IADC3, and the two-step IADC2. For OSR = 128, an IADC2 can achieve 85 dB, and an IADC3 can achieve 117 dB with a −6 dBFS input signal. Reusing the hardware of an IADC2 in a two-step operation, results in $SQNR_{21} \approx 110$ dB, 27 dB higher than SQNR$_2$. Also note that the IADC3 will be overloaded by a −6 dBFS input signal unless a high-resolution internal quantizer is used. A conventional 2-1 MASH modulator can mitigate the stability issue, but it requires three opamps to achieve third-order noise-shaping performance.

In the proposed two-step operation, the first-step IADC2 is operated only for 2/3 of the total conversion time. The SQNR loss is compensated by the second-step IADC1 operation. The energy efficiency is thereby improved significantly. The higher the OSR, the more significant the resulting SQNR improvement. The extra circuit cost is low: only an additional timing control is needed to switch the hardware between the two steps. The circuit configuration is much less complex than previously reported hardware-sharing extended-counting schemes.

5.3.2 Switched-Capacitor Circuitry

When the ADC is implemented in a 65 nm technology, the leakage current of the 1–V core MOS devices degrades the performance of a SC circuit operated at a low sampling frequency. However, the leakage current of the 2.5 V I/O devices is only 2 pA μm^{-1}, which is low enough even for a high-resolution ADC. Hence, 2.5 V I/O devices were used here to implement the prototype IADC.

The SC circuit implementation of the proposed two-step IADC's modulator is shown in Figure 5.21. Single-ended equivalent circuits are shown for simplicity, but the actual implementation is fully differential. A conventional resistor string was used to generate the five-level reference voltages $V_{R, i}$ for all comparators. To operate the 2.5-V MOS devices with a 1.2-V power supply, the charge pump circuits shown in Figure 5.21c were used to double the NMOS gate voltages of the sampling CMOS switches. The I/O devices do not suffer from gate and junction overdrive when operated at 2.5 V, and no extra transistors were needed to improve their reliability.

During the first step lasting $M_1 = 128$ clock periods, as shown in Figure 5.21a, the gray-scaled paths are not enabled, and the circuit is working as a conventional IADC2. To achieve a 100 dB SNR, the input sampling capacitor of the INT1 is designed to be 8 pF from kT/C thermal noise consideration. During the second step, for $M_2 = 64$ clock periods, as shown in Figure 5.21b the two-phase clocks S_1 and S_2 are disabled, and X_1 and X_2 establish different input paths reconfiguring the circuit as a first-order modulator. (In the SC circuitry used, it is simple to multiplex the different paths and to perform reconfiguration.) The INT2, which is now acting as a hold amplifier, drives the INT1's sampling capacitors. The input sampling capacitors of INT1 can, therefore, be reduced from 8 to 0.4 pF, to ease the loading of INT2.

Since in our circuit, the signal BW is 1 to 250 Hz, it is sensitive to flicker noise. The first opamp's in-band flicker noise is hence mitigated by chopping, at half of the 96 kHz sampling frequency. The signal is chopped during the middle of integrator sampling phase. The input chopping switches are turned off slightly before the output chopping switches, in order to reduce the signal-dependent charge injection from the output chopping switches. Careful layout techniques were employed to make sure that the in-band residual noise caused by chopper nonidealities is low.

In the proposed two-step IADC (Figure 5.19), the bit streams of each step are also separately accumulated and decimated. It has the same advantage as the MASH IADC: the digital circuitry providing $(1 - z^{-1})$ is no longer needed, and the opamp

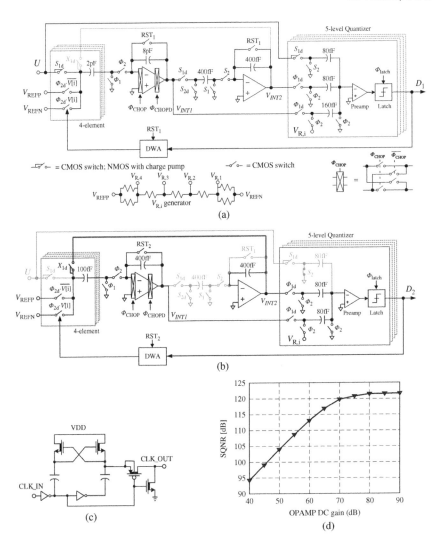

Figure 5.21 The equivalent single-ended switched-capacitor circuits implementation of the two-step IADC's modulator. (a) First step. (b) Second step. (c) Voltage doubler. (d) Simulated SQNR versus opamp gain.

gain is much relaxed. Figure 5.21d shows the simulated SQNR versus the opamp DC gain. The SQNR begins to degrade only when the opamp DC gain falls below 70 dB. The required opamp gain for the proposed two-step IADC is therefore quite low, even for very high resolution conversion. Although the loop gain in a third-order system can further relax the opamp gain, a second-order system with moderate relaxation can also save cost, and thus improve efficiency.

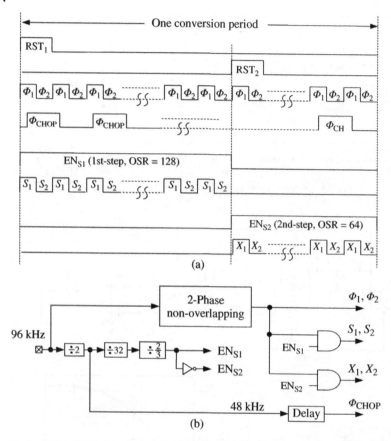

Figure 5.22 (a) The detailed timing diagrams for the two-step IADC2. (b) The circuit for timing control and two-phase non-overlapping clocks.

The detailed timing diagram for the SC circuitry is shown in Figure 5.22a. An external 96 kHz clock is used to generate the reset signals RST_1, RST_2 and the control signals EN_{S1}, EN_{S2}. The two-phase non-overlapping clock phases Φ_1 and Φ_2 at 96 kHz are used during both steps, while the S_1, S_2 and X_1, X_2 two-phase clock signals are specifically for the first and second steps, respectively. Bottom-plate sampling is used to mitigate the switches' nonidealities. The delayed versions of the two-phase clock signals Φ_{1d}, Φ_{2d}, S_{1d}, S_{2d}, X_{1d}, X_{2d}, and Φ_{CHOPD} are omitted for simplicity. The simplified circuit used to generate the control timing and two-phase clocks is shown in Figure 5.22b. It uses only frequency dividers and simple logic circuits, and hence, it is simple and does not need a complicated state machine to generate the timing controls.

For a total OSR of 192, the two-step IADC2 can ideally achieve 120 dB SQNR for a −6 dBFS input amplitude, which is adequate for a 100 dB SNR ADC. For comparison, a single-loop IADC2 with OSR = 128 and OSR = 192 can achieve only 84 dB and 91 dB SQNR, respectively. Increasing the OSR of an IADC2 from 128 to 192 can give only a 7 dB SNQR improvement, while increasing the order of the noise-shaping by 1 can improve the SQNR significantly, by 30 dB. The power penalty for the additional conversion time of 64 clock periods and for the extra control circuitry is small. By just enabling and disabling the control clocks of the switched capacitor circuit, a simple and low-cost operation results.

The digital decimation filter shown in Figure 5.19 is not implemented on the chip; the modulator's bit streams were post-processed using MATLAB. The detailed circuit design of the other building blocks can be found in Ref. [3].

5.3.3 Measured Performance

Defining the differential reference 2.4 V_{PP} as 0 dBFS, the measured spectra for a differential 100 μV_{PP} (−87.6 dBFS), 170 Hz, sine-wave input signal are shown in Figure 5.23a. The spectra obtained after the first step (OSR = 128) and the second step (OSR = 64) are plotted, showing that the second step enhances the SNDR by 10.3 dB. Figure 5.23b shows the measured spectra for a 2.2 V_{PP} (−0.76 dBFS) 17-Hz sine-wave input. The measured SNDR is 84.7 dB for the first step and 91 dB for two-step operation. Harmonic distortion limits the SNDR for such large signals, and the two-step operation enhances the SNDR by only 7 dB. Figure 5.24a shows the spectra with the digital-to-analog converter data-weighted average (DCA DWA) turned on and off. The measured SNDR with a 1 V_{PP} (−7.6 dBFS) input

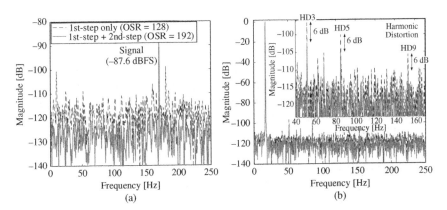

Figure 5.23 Measured spectra. The dotted line is for the first step only (IADC2, OSR = 128). The solid line is for the two-step IADC with OSR = 192. (a) 100 μV_{PP} (−87.6 dBFS) input amplitude. (b) 2.2 V_{PP} (−0.76 dBFS) input amplitude.

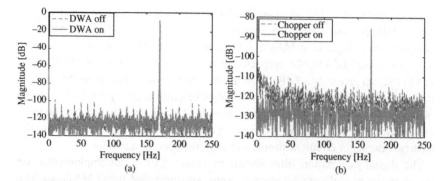

Figure 5.24 Measured spectra. (a) DWA turned on versus off with 1 V_{pp} (−6.8 dBFS) input amplitude. (b) Chopper on versus off with 100 μV_{pp} input amplitude.

amplitude is 84.6 dB (with the DWA on) and 75.8 dB (DWA off). Nevertheless, the in-band flicker noise degrades the SNR performance significantly. Figure 5.24b shows the measured spectra with the chopper turned on and off. The chopper stabilization reduces the in-band flicker noise by 11 dB.

The ADC achieves a DR of 99.8 dB and a peak SNDR of 91 dB with a BW from 1 to 250 Hz, consuming only 10.7 μW. Table 5.1 shows a performance summary and comparison with recent state-of-the-art single-loop IADCs [1, 5], as well as with hybrid IADCs using extended-counting schemes [8–10]. The Walden FoM$_W$ and Schreier FoM$_S$ were also calculated. For this device, FoM$_W$ = 0.76 pJ/conv.-step and FoM$_S$ = 173.5 dB were found, both among the best reported results.

5.3.4 A Two-Step Third-Order IADC

Generally, if the two-step architecture is applied to an Nth-order IADC, its performance will be boosted up to nearly that of a $(2N-1)^{th}$-order one. By using an IADC3 in a two-step operation, as shown in Figure 5.25a, an IADC3 performs the first step, and then it is reconfigured as an IADC2 for the second step, as shown in Figure 5.25b. Figure 5.25c plots the simulated SQNR of the two-step IADC3 versus the total OSR. Compared to a single-loop IADC3 and IADC2, the two-step IADC3's performance is indeed nearly equal to that of an IADC5. The optimal SQNR can be achieved for an OSR ratio of 3 : 2, which can be derived as in Ref. [3]. However, the improvement is more significant when the OSR is higher.

In Ref. [11], an algorithmic IADC was proposed with similar two-step operation. However, it requires an extra sample-and-hold stage, and also an additional clock phase to feedback the residue voltage. This complicates the circuit implementation. In our proposed two-step IADC, neither an additional phase nor an extra active component is needed. All components are reused to accomplish higher SQNR performance, and the power consumption remains the same.

Table 5.1 Performance summary and comparison.

Parameters	[3] This work	[7] ESSCIRC '13	[8] TCAS-I '10	[9] JSSC '10	[10] ISSCC '13	[1] JSSC '06
Architecture	IADC2 + IADC1	10b SAR + IADC1	IADC2 + 10b cyclic	IADC2 + 11b SAR	Single-loop IADC2	Single-loop IADC3
Process	65 nm (2.5 V MOS)	0.6 μm	0.18 μm	0.18 μm	0.16 μm	0.6 μm
Area (mm²)	0.20	1.64	0.50	3.5	0.45	2.08
VDD (V)	1.2	3.3	2	1.8	1	3
Sampling freq.	96 kHz	5 MHz	115 MHz	45.2 MHz	750 kHz	30.7 kHz
OSR	192	256	5	45	80	512
Input range	$2.2\,V_{pp}$	$2\,V_{pp}$	$3.6\,V_{pp}$	$2\,V_{pp}$	$0.7\,V_{pp}$	$6\,V_{pp}$
Dyn. range (dB)	99.8	84.6	73	90.1	81.9	120
Peak SNDR (dB)	90.8	70.7	72	86.3	81.9	120
Bandwidth (Hz)	250 Hz	9.75 kHz	11.5 MHz	500 kHz	667 Hz	7.5 Hz
Power	10.7 μW	64 μW	48 mW	38.1 mW	20 μW	300 μW
FoM$_W$[b] (pJ/conv.)	0.76	1.17	1.02	1.46	1.48	24.46
FoM$_{S,DR}$[a] (dB)	173.5	166.4	156.8	161.3	157.1	164.0

a) FoM$_{S,DR}$ = DR + 10 log$_{10}$ (BW/Power).
b) FoM$_W$ = Power/($2^{(SNDR-1.76)/6.02} \cdot 2 \cdot$ BW).

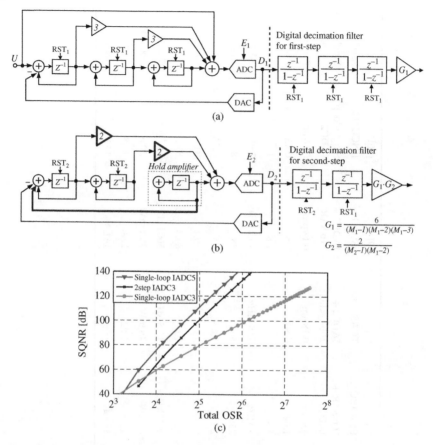

Figure 5.25 The IADC3 in two-step operation. (a) First step. (b) Second step. (c) Simulated SQNR versus single-loop IADC5 and IADC3.

5.3.5 Conclusion

To further improve the energy efficiency, we proposed multistep operation for high-resolution ADCs for use in integrated sensor interface circuits. For example, the components of an Nth-order IADC can be reused to quantize the residue voltage in a second-step operation, resulting in noise-shaping performance close to that of an IADC of order $(2N-1)$. The extra cost is only simple added timing control circuits. Moreover, the required opamp gains can be as low as 60 dB even for 100 dB SNR. The principle can be extended to three- and higher-step operations.

A design example of a two-step IADC2 was demonstrated. The ADC was fabricated using 2.5 V I/O MOS devices in a 65-nm technology and operated with a 1.2 V power supply. The measured performance showed a 100 dB DR and 91 dB maximum SNDR for a signal BW from 1 to 250 Hz. The device consumed only 10.7 µW. The measured Walden and Schreier FoMs were 0.76 pJ/conversion-step and 173.5 dB, respectively, among the best-published IADC FoMs. The active area is 0.2 mm^2, which is the smallest among published designs. The results verify that the proposed two-step IADC is a very area- and energy-efficient solution for integrated sensor systems.

5.4 A Hybrid Continuous-Time Incremental and SAR Two-Step ADC with 90.5 dB Dynamic Range Over 1 MHz Bandwidth

In a discrete-time IADC, the nonlinearity and charge injection of the sampling switch degrades the performance, and power-hungry opamps are needed to provide fast and accurate settling for the switched capacitor circuits. These limitations may prevent the use of DT-IADCs in high resolution and wide BW applications. CT-IADC overcome these problems by removing the sampling switches, and CT integrators allow relaxed specifications for the amplifiers' slew rate settling accuracy to save power. Hence, CT-IADCs enable higher resolution, faster conversion speed with lower power consumption.

Single-stage IADCs achieve large SQNRs by using high OSR, high-order loop, and high-resolution quantizer. However, achieving these becomes problematic for wide BW signals.

Hence, a two-stage scheme was chosen for the device described in this project [4]. The block diagram of the system is shown in Figure 5.26.

The structure has the following features:

1. It uses a CT first stage, which avoids the linearity issues with the input sampling switches encountered in DT-IADC, and reduces the bias currents needed in the integrators.
2. It uses feedforward path between the input and the quantizer, which reduces the nonlinear distortion in the loop filter [12].
3. It uses the robust MASH ΔΣ ADC configuration [11] which reduces the leakage of the first-stage quantization error E_1 to the output caused by imperfections in the first-stage STF, by transferring and canceling the shaped noise NTF_1 E_1, rather than E_1.
4. It introduces an analog delay between the CT first stage and the DT second one, to synchronize the operations of the two stages.
5. It uses an FIR filter $H_1(z)$ derived by the zero-order hold (ZOH) mapping between the analog and digital variables s and z to match the transfer functions of the two paths: Path 1 and Path 2 between the quantizer output and the ADC

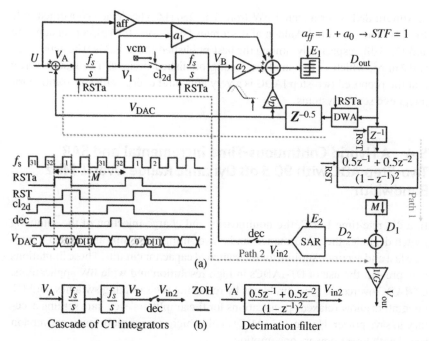

Figure 5.26 (a) System diagram. (b) CT to DT transformation.

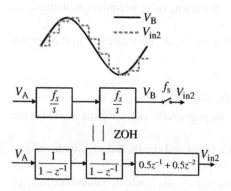

Figure 5.27 The ZOH *s–z* mapping.

output. This ensures the accurate cancelation of $NTF_1 E_1$. The transformation is illustrated in Figure 5.27. The circuit uses a half-period excess loop delay (ELD) correction (a_0).

6. In the operation of the IADC, the weight of the first output is the highest in the digital filter. Hence, the delay in the feedback path introduced by the parasitic capacitances of the DAC after the last conversion period causes would introduce

a large error. To prevent this, the last period is used to reset all memories in the circuit.

7. The CT integrator input is disconnected in the last cycle to let the integrator hold the residual error and give more time for SAR to sample.

8. The CT integrator input is disconnected in the last cycle to let the integrator hold the residual error and give more time for SAR to sample.

Figure 5.28 shows the simplified circuit diagram of the IADC. The shaded area indicated the opamp assistants used to reduce the slewing distortion of the first opamp. Its effectiveness is illustrated by the transient responses shown.

The proposed 2-step architecture has a 16b resolution and 1 MHz BW. During the first step, it functions as a 2nd-order CT-IADC clocked at f_s = 64MHz with OSR = 32, and provides an SQNR = 75 dB. Since the NTF_1 does not contribute to the overall SQNR of the ADC, a moderate out-of-band gain (OBG) = 2.2 was chosen to reduce the inter-symbol interference (ISI) and jitter noise of DAC by having less DAC transitions.

Using a 4-bit internal flash quantizer allows the DAC to tolerate a 2 ps rms jitter noise and provides more than 96 dB SNR. A dual-return-to-zero (DRZ) DAC [13] is used, which converts the dynamic ISI errors of the DAC cells to a signal-independent static error, which can be canceled by the DWA used.

After reusing the integrators following the reset, each integrator has a step output, and the opamp output swing increases significantly, causing also a large jump at its virtual ground. As a result, the DRZ DAC linearity becomes much worse in the first cycle, and the performance of CT-IADC deteriorated due to the large weight of the first output sample. The assistant gm blocks and the non-return-to-zero (NRZ) DAC are introduced at the INT1 output to track the input and DRZ DAC output currents. This relaxes the output current requirement of the first opamp [14]. It also reduces the noise at the virtual ground, improving the DAC linearity. The finite unity-gain-bandwidth (UGB) of the opamp introduces a gain error and an extra pole into the integrator's transfer function [15]. The gain error can be compensated at the 2nd stage SAR ADC's output. The extra pole effect is negligible when the UGB is higher than 4 fs. With the 4-bit internal quantizer, the 8-bit 2nd stage SAR ADC increases the ENOB of the ADC by 4 bits, and the SQNR is increased by 24 dB. To allow the use of smaller sampling capacitors and fewer reference voltages, a two-capacitor SAR ADC architecture [16] was implemented. Figure 5.29 shows the circuit of the SAR ADC, and Figure 5.30 illustrates the spectra after Step 1 and Step 2.

The prototype of the two-step ADC was implemented in a 180 nm CMOS process, occupying an active area of 3.11 mm^2. It consumes 29.3 mW from a 1.8 V supply. The die micrograph is shown in Figure 5.31.

Figure 5.32 illustrates the measured SNDR versus sine wave input amplitude. The measured DR was 90.5 dB for the two-step ADC. Since the first stage CT ADC

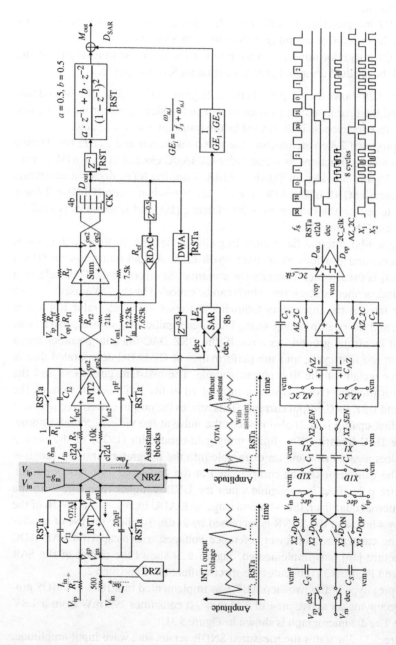

Figure 5.28 The circuit diagram of the hybrid IADC.

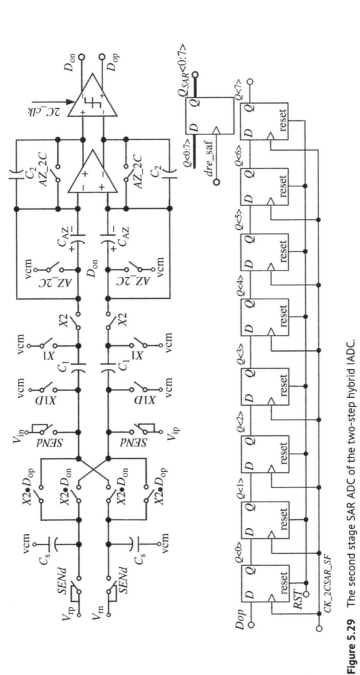

Figure 5.29 The second stage SAR ADC of the two-step hybrid IADC.

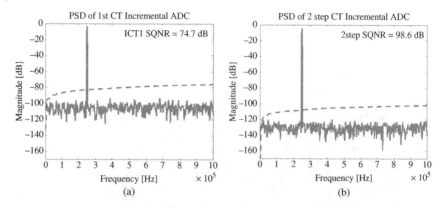

Figure 5.30 (a) Output spectrum after Step 1; (b) output spectrum after Step 2.

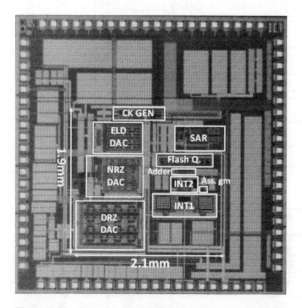

Figure 5.31 Die micrograph of the hybrid two-step IADC.

input does not require a sampling switch, it is not sensitive to switch nonlinearity. Hence, a high-frequency input signal was used to perform the tests.

Figure 5.33a shows the measured PSD at the peak SNDR for a 990 kHz input signal. The CT-IADC achieves a 72 dB SNDR and 76.8 dB spurious-free dynamic range (SFDR). This improves to 80.5 dB SNDR and 86 dB SFDR for the two-step ADC. When the input signal amplitude becomes large, the dynamic errors increase, making the noise floor of the 2-step ADC move down by 9 dB from that

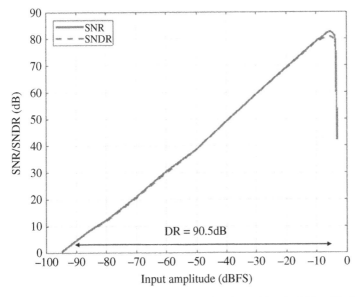

Figure 5.32 The measured SNR and SNDR at $f_{clk} = 64\,MHz$, OSR $= 32, f_{in} = 990$.

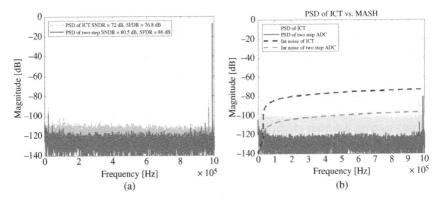

Figure 5.33 Measured output spectra: (a) with full-scale input; (b) with $-80\,dBFS$ input at 990 kHz.

of the CT-IADC. The measured power density spectrum (PSD) for a small input signal is shown in Figure 5.33b. The PSD is now dominated by the thermal noise and the quantization noise. The overall ADC noise floor is around 23 dB lower than for the CT-IADC alone.

Table 5.2 presents the performance summary and comparison with state-of-art CT-IADCs. The described device achieves 90.5 dB DR and 165.5 dB FoMs for a

Table 5.2 Performance summary and comparison with state-of-art CT-IADCs.

	[4] This work	[17]	[14]	[18]	[19]
Architecture	CT-IADC SAR	CT-IADC	CT-IADC	CT-IADC	CT-IADC
Technology (nm)	180	180	180	180	28
Area (mm²)	3.99	0.175	0.35	0.337	0.125
Supply (V)	1.8	3	1.8	1.8/1.2	0.9
Power (mW)	29.3	1.27	0.055	0.0348	3.6
Fs (MHz)	64	32	3.048	0.32	120
Fnyq (Ks/s)	2000	200	12	8	2000
SNDR (dB)	80.5	83	85.1	75.9	81.2/78.4
SFDR (dB)	85	94.3	97	88.1	97/94.8
DR (dB)	90.5	91.5	88[b]	85.5	89
FoM$_{S,DR}$[a] (dB) @ f$_{in}$	165.5 @ 990 KHZ	170.4 @11 KHz	168.4 @6 KHz	168.4 @175 Hz	173.4 @110 KHz

a) FoM$_{S,DR}$ = DR + 10 · log$_{10}$ (BW/Power).
b) Estimated from given plots.

990 kHz input, which are comparable to the numbers for state-of-art IADCs with 1 MHz BW. The results verify the usefulness of CT-IADCs for high-speed and low-noise applications.

5.5 A Multistage Multistep IADC

In this section, a design example of a multistage IADC in multistep operation is introduced. Based on the design of a MASH 1-1 IADC in Section 4.1, the analysis and design of a MASH 2-1 IADC is introduced first. The MASH 2-1 IADC operated in two steps is proposed to fine-quantize the residue voltage of the MASH 2-1 IADC. Circuit design and implementation are illustrated next. Measured performance and comparison of the hardware prototype in 0.18 μm technology are described finally.

5.5.1 Design of a MASH 2-1 IADC

By cascading an IADC2 and an IADC1, the MASH technique can be applied to IADC operation. Figure 5.34a shows an example of MASH 2-1 IADC (IADC2-1), and its digital filter and simplified timing control are depicted in Figure 5.34b,c, respectively. It is easier to derive the operation of FIR by time-domain difference

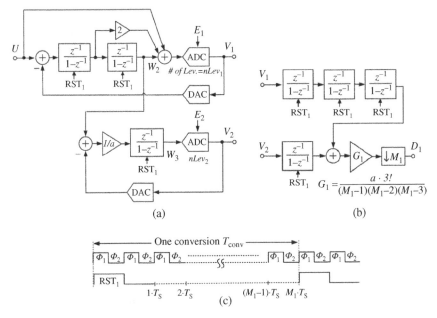

Figure 5.34 (a) An example of a MASH 2-1 IADC. The interstage gain G_i is not needed when SQNR is high enough. (b) The digital filter. (c) The control timing, including two-phase clock and reset pulse.

equations. The conversion starts with cleaning up the data in analog and digital circuitry by a reset pulse RST_1. Defining the OSR as the number of oversampling clock periods in one conversion time period (T_{CONV}), the INT2's output node $W_2[i]$ at the end of T_{CONV} can be described as

$$W_2[M_1] = \sum_{j=1}^{M_1-1} \sum_{i=1}^{j-1} U[i] - \sum_{j=1}^{M_1-1} \sum_{i=1}^{j-1} V_1[i] \tag{5.17}$$

Correlating the input and output at the quantizer of the second IADC1, the closed-loop design equation can be finished.

$$\frac{1}{a}\left[\sum_{i=1}^{M_1-1} W_2[i] - \sum_{i=1}^{M_1-1} V_2[i]\right] + E_2[M_1] = V_2[M_1] \tag{5.18}$$

By using Eqs. (5.17) and (5.18), the difference equation of the multistage design can be described

$$\sum_{k=1}^{M_1-1} \sum_{j=1}^{k-1} \sum_{i=1}^{j-1} U[i] + a \cdot E_2[M_1] \approx \sum_{k=1}^{M_1-1} \sum_{j=1}^{k-1} \sum_{i=1}^{j-1} V_1[i] + \sum_{i=1}^{M_1} V_2[i] \tag{5.19}$$

The input signal is averaged by triple integration-and-dump and the average $\overline{U}$ can be calculated as

$$\overline{U} = \frac{3!}{(M_1 - 1)(M_1 - 2)(M_1 - 3)} \sum_{k=1}^{M_1-1} \sum_{j=1}^{k-1} \sum_{i=1}^{j-1} U[i] \qquad (5.20)$$

The input signal now can be estimated as

$$\overline{U} + \frac{a \cdot 3!}{(M_1 - 1)(M_1 - 2)(M_1 - 3)} E_2[M_1] = \frac{a \cdot 3!}{(M_1 - 1)(M_1 - 2)(M_1 - 3)}$$
$$\times \left[\sum_{k=1}^{M_1-1} \sum_{j=1}^{k-1} \sum_{i=1}^{j-1} V_1[i] + \sum_{i=1}^{M_1} V_2[i] \right] \qquad (5.21)$$

Ideally, the combination of the two loop filters cancels the E_1, and only the E_2 of the IADC1 appears in the multistage to contribute the quantization noise. The effective LSB error of the IADC2-1 is

$$E_{2-1} = \frac{V_{FS}}{nLev_2 - 1} \cdot \frac{a \cdot 3!}{(M_1 - 1)(M_1 - 2)(M_1 - 3)} \qquad (5.22)$$

where the V_{FS} is full-scale voltage and $nLev_2$ is the number of levels of the internal quantizer in the IADC1. The SQNR of IADC2-1 can thus be estimated

$$SQNR_{2-1} = 10 \log \left(\frac{V_{FS}^2/8}{E_{2-1}^2/12} \right) \approx 20 \log \left(\frac{(nLev_2 - 1) \cdot M_1^3}{a \cdot 3!} \right) \qquad (5.23)$$

The SQNR can be designed higher by increasing the $nLev_2$ of the second quantizer. The interstage gain G_1 can be added to increase the SQNR. When the SQNR is adequate, no additional interstage gain is needed ($G_1 = 1$), and 1-bit quantizer can be used in the second loop ($nLev_2 = 2$). The coefficient $1/a$ in the second-loop IADC1, for example, $1/a = 1/3$, can be incorporated to scale the DR and the 1-bit modulator can retain a wide non-overloaded range. In the following analysis, we assume that $G_1 = 1$.

To reconstruct the input signal, the oversampled bit streams V_1 and V_2 of each loop are accumulated concurrently and decimated at Nyquist-rate as a digital sample D according to the RHS of Eq. (5.21). The customized reconstruction filter, depicted in Figure 5.34b, can be implemented by simply cascading counters.

5.5.2 An IADC2-1 Versus a MASH 2-1 $\Delta\Sigma$ ADC

In the conventional MASH modulators, an additional error cancellation logic (ECL) $H_{ECL}(z)$ must be incorporated before the decimation filter [20]

$$H_{ECL}(z) = V_1 \cdot z^{-1} + V_2 \cdot (1 - z^{-1})^2 \qquad (5.24)$$

Figure 5.35 illustrated the digital signal processing of a MASH 2-1 $\Delta\Sigma$ ADC. An ECL in the digital domain cancels the E_1 of the first loop and the error does not

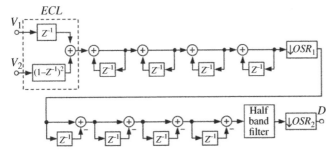

Figure 5.35 The digital ECL logic and decimation filter in a conventional MASH 2-1 $\Delta\Sigma$ ADC with total $OSR = OSR_1 \cdot OSR_2$. The latency introduced can be approximated as $4 \cdot OSR \cdot T_S = 4 \cdot T_{CONV}$. The latency can be longer if a higher-order HBF is used to filter out-of-band noise. Source: Adapted from Lyons [21].

appear in the MASH 2-1 loops. In reality, mismatch between digital and analog noise-shaping function $(1 - z^{-1})^2$ causes leakage of quantization noise, and thus the SQNR is degraded. In Ref. [20], the opamp dc gain is designed as high as 96 dB to achieve very high resolution. While operated as an IADC, the ECL is no longer needed and the hardware requirement can therefore be relaxed. More nonidealities will be described in Section 5.5.5.

The decimation filter in the conventional $\Delta\Sigma$ ADCs is usually implemented by a Hogenauer sinc filter $[(1 - z^{-M1})/(1 - z^{-1})]^4$ as the front-end decimator [22], and one or more half-band filters (HBF) continue the rest of decimation. If the oversampled bit stream is directly downsampled to Nyquist-rate using only a sinc filter without an HBF, the latency is $4 \cdot OSR \cdot T_S = 4 \cdot T_{CONV}$. Multistage decimation by using an extra higher-order HBF in Figure 5.35 can filter out more out-of-band noise, but it can introduce longer latency [21]. Comparing the digital filter in Figure 5.35 against Figure 5.34b, we find that the digital filter of IADC2-1 is much simpler. Furthermore, the latency of IADC2-1 is only one T_{CONV}, which is much shorter than that of the $\Delta\Sigma$ counterparts. These design benefits make the IADC2-1 much more favorable.

5.5.3 Noise Consideration for Higher-Order IADCs

When a higher SQNR is required, it is straightforward to cascade one more stage [23] to increase the order of noise shaping. Higher-order noise shaping can reduce the OSR and relax the speed limit. Unfortunately, a higher-order IADC suffers from higher noise penalty [23, 24] and it does not improve the energy efficiency. In a $\Delta\Sigma$ ADC, the input-referred noise power is $P_{n,in} = \overline{v_{n,s}^2}/M_1$, where $\overline{v_{n,s}^2}$ is the input noise power of each sample. As the same higher-order modulator is used as an IADC, every conversion has a time-dependent weight factor w_i that is correlated to each input sample. When the total noise power is referred to the IADC's input, it becomes

$$P_{n,in} = \sum_{i=1}^{M_1} w_i^2 \overline{v_{n,s}^2} = \overline{v_{n,s}^2} \cdot \sum_{i=1}^{M_1} w_i^2 \tag{5.25}$$

For an IADC1, all the weight factors $w_1 = w_2 = \ldots = w_{M1} = 1/M_1$ and w_i is an equal weight for all M_1 samples. Then the IADC1 sampled thermal noise is still $\overline{v_{n,s}^2}/M_1$. In an IADC2 with two CoI as the digital reconstruction filter, the w_i becomes

$$w_i = \frac{2}{M_1(M_1+1)} \cdot i \approx \frac{2}{M_1^2} \cdot i \tag{5.26}$$

The IADC2's input-referred noise power now is proportional to

$$\overline{v_{n,s}^2} \cdot \sum_{i=1}^{M} w_i^2 = \overline{v_{n,s}^2} \cdot \left(\frac{2}{M_1^2}\right)^2 \cdot (1^2 + 2^2 + 3^2 + \cdots + M_1^2) \approx \frac{4}{3} \cdot \frac{v_{n,s}^2}{M_1} \tag{5.27}$$

when M_1 is large enough, $\sum w_i^2 \approx 1.33/M_1$ for an IADC2. Similar derivation can be repeated for higher-order IADCs. $\sum w_i^2$ approaches $1.8/M_1$ for third-order IADCs, and it is $2.29/M_1$ for fourth-order ones. The IADC2-1 has the third-order characteristics and the noise power is 1.8 times that of a $\Delta\Sigma$ ADC. To maintain the SNR performance, the input capacitance should be sized larger, and the opamps dissipate extra power to drive heavier loads. Therefore, higher-order loop filters operated as IADCs are less energy-efficient than $\Delta\Sigma$ ADCs.

In Ref. [18], two CT-IADC2s are cascaded, and a sample-hold (S/H) circuit is added to interface the two IADC2 loops. The S/H can avoid the noise power of two IADC2 loops building up as a fourth-order one. The equivalent input-referred noise is dominated by the first IADC2, while the noise contribution by interstage S/H and the second IADC2 is negligible. However, the additional S/H to hold the residue voltage is not necessary [23] when two IADC loops are cascaded. Besides, this additional S/H consumes extra power.

5.5.4 The Proposed Multistage Multistep IADC

To satisfy the SQNR requirement of an IADC, either the noise-shaping order or the OSR must be designed higher. Increasing one more noise-shaping order can be more effective than doubling the OSR [3, 25]. For example, to have SQNR sufficiently higher than 100 dB operated at OSR = 32, at least a fourth-order circuit is required. To mitigate the narrower non-overloaded range and avoid extra thermal noise penalty introduced in the higher-order circuits, the multistage IADC2-1 is proposed in multistep operation by reusing the same hardware of IADC2-1 [26, 27].

The multistep operation starts with the normal IADC2-1 described in Section 5.5.1 as the first step for M_1 clock periods (OSR = M_1). When the IADC2-1 operation is finished (time index = M_1), the third integrator's output voltage $W_3[M_1]$ is

$$V_{RES} = W_3[M_1] = \frac{1}{a}\left[\sum_{k=1}^{M_1-1}\sum_{j=1}^{k-1}\sum_{i=1}^{j-1}U[i] - \left(\sum_{k=1}^{M_1-1}\sum_{j=1}^{k-1}\sum_{i=1}^{j-1}V_1[i] + \sum_{i=1}^{M_1-1}V_2[i]\right)\right]$$

(5.28)

Examining Eq. (5.28) versus (5.19), we can find out that $a_2 \cdot E_2[M_1]$ in the RHS of Eq. (5.19) is the quantization residue voltage, which is accumulated and stored at the third integrator at W_3 in Eq. (5.28). The residue (V_{RES}) is ready to be further quantized, and the IADC2-1 can now be reused and reconfigured to continue fine quantization. Figure 5.36 illustrates the z-domain circuit reconfiguration. V_{RES} is held constantly and sent into the IADC2 loop. The original IADC2-1 is reworked as an IADC2 for another M_2 clock periods.

The detailed derivation of the IADC2 design equations can be found in Refs. [3, 25]. At the end of the second-step IADC2 operation, the fine quantization of V_{RES} can be described as

$$\sum_{j=1}^{M_2}\sum_{i=1}^{j}V_{RES} + E_1[M_2] = \sum_{j=1}^{M_2}\sum_{i=1}^{j}V_1[i]$$

(5.29)

$$V_{RES} + \frac{2\cdot E_1[M_2]}{M_2(M_2+1)} = \frac{2}{M_2(M_2+1)}\sum_{j=1}^{M_2}\sum_{i=1}^{j}V_1[i]$$

(5.30)

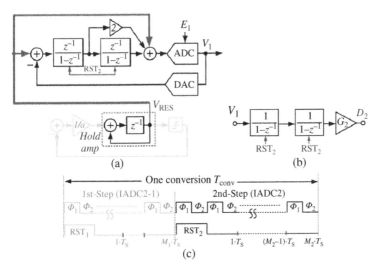

(a)

(b)

(c)

Figure 5.36 (a) The second-step operation of the proposed two-step MASH2-1. It is reconfigured as an IADC2. (b) The digital filter. (c) Control timing.

By inspecting Eq. (5.30), it is straightforward to see that the V_{RES} is fine-quantized by additional second-order noise shaping. The complete two-step data conversion can be derived by using Eqs. (5.28) and (5.30), with the digital scaling factors G_1 and G_2 defined in Eq. (5.31).

$$G_1 = \frac{3!}{(M_1 - 1)(M_1 - 2)(M_1 - 3)} \quad G_2 = \frac{2}{M_2(M_2 + 1)} \tag{5.31}$$

$$\frac{1}{a}\sum_{k=1}^{M_1-1}\sum_{j=1}^{k-1}\sum_{i=1}^{j-1} U[i] + G_2 \cdot E_1[M_2] = \frac{1}{a}\left(\sum_{k=1}^{M_1-1}\sum_{j=1}^{k-1}\sum_{i=1}^{j-1} V_1[i] + \sum_{i=1}^{M_1-1} V_2[i]\right)$$

$$+ G_2 \sum_{j=1}^{M_2}\sum_{i=1}^{j} V_1[i] \tag{5.32}$$

By using the average $\overline{U}$ in Eq. (5.20) and rewriting Eq. (5.32), we can finish the design.

$$\overline{U} + a \cdot G_1 \cdot G_2 \cdot E_1[M_2] = G_1 \left[\sum_{k=1}^{M_1-1}\sum_{j=1}^{k-1}\sum_{i=1}^{j-1} V_1[i] + \sum_{i=1}^{M_1-1} V_2[i]\right]$$

$$+ a \cdot G_1 \cdot G_2 \sum_{j=1}^{M_2}\sum_{i=1}^{j} V_1[i] \tag{5.33}$$

The digital filter to reconstruct the two steps can be designed by the RHS of Eq. (5.33). The $a_2 \cdot G_1 \cdot G_2 \cdot E_1[M_2]$ in the left-hand-side of Eq. (5.33) stands for the equivalent of one LSB error of the proposed multistep IADC2-1 (IADC2-1-2)

$$E_{2-1-2} = a \cdot G_1 \cdot G_2 \cdot E_1[M_2] \approx \frac{V_{FS}}{nLev_1 - 1} \cdot \frac{a \cdot 3!}{M_1^3} \cdot \frac{2}{M_2^2} \tag{5.34}$$

Ideally, only the error $E_1[M_2]$ of the second step appears in the multistage multistep loops within the total OSR ($M = M_1 + M_2$). In Eq. (5.34), one LSB error $V_{FS}/(nLev_1 - 1)$ of the internal quantizer is first third-order shaped by the IADC2-1 in M_1 clock periods, and then second-order shaped by an IADC2 in M_2 clock periods. The SQNR relative to the input full-scale voltage can be calculated.

$$SQNR_{2-1-2} \approx 20\log\left(\frac{(nLev_1 - 1) \cdot M_1^3 \cdot M_2^2}{a \cdot 3! \cdot 2!}\right) \tag{5.35}$$

Within the same T_{CONV}, the effective OSR of the proposed IADC2-1-2 is lower than that of a true fifth-order IADC (IADC5). It is reasonable that the $SQNR_{2-1-2}$ is inferior to a true IADC5's SQNR. By selecting the OSR M_1 and M_2 of each step properly, the $SQNR_{2-1-2}$ can be optimized and the IADC2-1-2 nearly equals to that

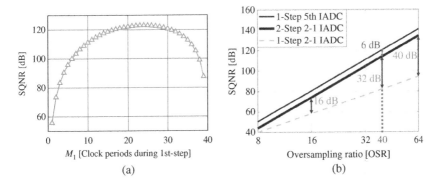

Figure 5.37 (a) SQNR versus the OSR of the first step (M_1) within a total OSR $= 40 = M_1 + M_2$. The optimal ratio is $M_1 : M_2 = 3 : 2$. (b) Simulated SQNR of the proposed IADC2-1-2 versus OSR and comparison with a true 5th-order IADC and an IADC2-1 in one step.

of an IADC5. By defining $M_1/M_2 = k_{12}$, we can obtain $M_1 = k_{12} \cdot M/(k_{12} + 1)$ and $M_2 = M/(k_{12} + 1)$. The $E_{2\text{-}1\text{-}2}$ in Eq. (5.34) can be described in terms of M and k_{12}.

$$E_{2-1-2} = \frac{V_{\text{FS}}}{nLev - 1} \cdot \frac{a \cdot 3! \cdot 2}{M^5} \cdot \frac{(k_{12} + 1)^5}{k_{12}^3} \tag{5.36}$$

when $k_{12} = M_1/M_2 = 3/2$ is selected, the $E_{2\text{-}1\text{-}2}$ is minimized and SQNR$_{2\text{-}1\text{-}2}$ has an optimal value. Figure 5.37a plots the SQNR$_{2\text{-}1\text{-}2}$ versus M_1 with a total OSR of 40, and it verifies the highest SQNR for $M_1 : M_2 = 24 : 16$. Nonetheless, the $SQNR_{2\text{-}1\text{-}2}$ is not sensitive to the optimal ratio, and there is a degree of freedom to choose the OSR of each step. When considering thermal noise, the M_1 can be longer and M_2 is shorter. The larger M_1 will lower $kT/C/M_1$, while the noise penalty factor remains almost the same. For example, the $M_1 : M_2$ can be changed to $32 : 8$ and the SQNR loss is only several dB.

Figure 5.37b compares the proposed IADC2-1-2 versus an IADC5 and an IADC2-1 with the same OSR, assuming that modulators have one-bit internal quantizer and do not overload when input 0 dBFS amplitude. Even though the effective OSR of the first-step IADC2-1 is reduced, the second step increases the noise shaping by two orders, which enhances the SQNR very significantly. The IADC2-1-2 achieves a SQNR of 114 dB, which is 32 dB higher than an IADC2-1. The total noise of an IADC2-1 is dominated by the quantization noise, resulting in a maximum SNR of only 82 dB. When considering thermal noise, it inevitably increases in the IADC2-1-2 due to the reduced effective OSR.

In Eqs. (5.34)–(5.35), the SQNR$_{2\text{-}1\text{-}2}$ is also dependent on $nLev_1$ of the first quantizer. Increasing the $nLev_1$ can improve both the SQNR$_{2\text{-}1\text{-}2}$ and stability. However, increasing $nLev_2$ does not help increase SQNR, but it can help to maintain a wide non-overloaded range wide. Since $nLev_1 = 5$ has been chosen and SQNR$_{2\text{-}1\text{-}2}$ of

114 dB is more than what is needed, using a one-bit quantizer ($nLev_2 = 2$) with 1/3 DR scaling only results in a few dB of SQNR loss. The non-overloaded range remains wide with minimal circuits.

In the same conversion, the total OSR of 40 is split between IADC2-1 and IADC2, resulting in an effective OSR in each step less than 40. With less OSR, the same fifth-order noise shaping results in lower SQNR. The maximum $SQNR_{2-1-2}$ can be estimated using Eq. (5.35) with $a = 1$. Under the same conditions, the SQNR of a true IADC5 that uses the entire OSR of 40 can be estimated by replacing $M_1 = 24$ and $M_2 = 16$ with 40 in Eq. (5.35). Detailed mathematical calculations show that the IADC2-1-2 is only 6 dB lower than that of an IADC5. When a true IADC5 is implemented, either a single-loop or a multistage, too much circuit overhead must be paid. The energy efficiency is degraded significantly and undesirable oversize silicon area is occupied. A true fifth-order circuit with five opamps and much more peripheral circuitry can rarely be found among published works. The digital filters in each step in Figure 5.34b and Figure 5.36b can be customized and the complete reconstruction filter is shown in Figure 5.38. CoIs for a true IADC5 are shown in Figure 5.38b for comparison with the digital filter for the IADC2-1-2. The IADC5 requires two more CoIs and uses more registers since the n-bit length of each integrator will be extended to $n + \log_2$ (OSR). The IADC2-1-2 requires two more adders, but fewer CoIs.

In Section 5.5.2, an IADC2-1 is compared to a MASH 2-1 modulator. Even though the IADC2-1-2 exhibits more thermal noise, it has less demanding hardware requirements than a conventional MASH $\Delta\Sigma$ modulator. It is challenging to reuse a MASH 2-1 modulator as a hybrid ADC for multistep quantization. The beauty that the IADCs possess inside is the inherent residue voltage generation [3, 25]. To fulfill the need for multistep quantization, the proposed IADC2-1-2 can make it easier. The fifth-order can be simply accomplished by reusing the same hardware of a third-order IADC2-1 with least excess circuit efforts.

To further compare multi-bit quantization with the noise-shaping order in Figure 5.36a, reconfiguring the circuits as a binary search ADC to perform

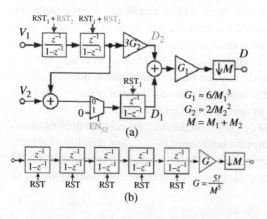

(a)

(b)

Figure 5.38 (a) The complete digital filter to reconstruct the proposed IADC2-1-2. (b) The CoIs as the digital filter for a true IADC5.

extended counting can be more energy efficient. For example, using a 10-bit SAR ADC [25, 28] or a cyclic ADC [29, 30] can increase the SQNR by 60 dB within additional 10 clock periods in the second step. The residue stored at W_3 by an amplifier in Figure 5.36a must drive the capacitance load of the 10-bit SAR ADC. Using a much smaller capacitor as the SAR DAC can mitigate the driving load of the hold amplifier. However, this approach comes with several design challenges and complexities. The use of very small unit capacitance poses difficulties in achieving good matching and ENOB. The interconnection and routing of 10-bit DAC introduce inevitable parasitic, leading to mismatches and gain errors. The parasitic capacitance at the summing node, located at the input of the comparator further contributes to gain errors, which degrade SQNR drastically. As an alternative, the residue holder can be reconfigured as a binary search cyclic ADC. Using either a binary capacitor array or a multiply-by-two cyclic amplification, zero-order extended counting suffers from gain error and coefficient mismatch which drastically degrade the SQNR. Moreover, the comparator's offset must be very low when circuits are reused as the binary counting phase [29]. Reusing IADC loop filters as binary search schemes introduces very challenging nonidealities. Not to mention that the circuit also becomes much more complicated.

Zoom ADCs are 0-L MASH and they can increase the effective quantizer's levels by utilizing SAR coarse quantization. The one-bit modulator loop filters only need to convert the residue, which has a very small amplitude. The two-step operation can only convert dc signals but the FoM is excellent [31, 32]. The dynamic zoom ADC [33] operates the coarse and fine ADCs concurrently, with more circuit complexity. The signal BW is limited by the coarse SAR resolution, and using more coarse bits increases the peripheral circuit drastically. The STF deviates from 1 and this high frequency peaking results in imperfect cancelation of the SAR quantization error [34]. In general, zoom ADCs achieve very excellent energy efficiency but with inevitable design complexity. The proposed IADC2-1-2 is much simpler, and its circuit implementation will be shown in Section 5.5.5. It has the potential to be a solution for merging these two schemes to enhance the energy efficiency.

5.5.5 Circuit Implementation

The SC circuits implementation of the IADC2-1 in the first-step operation is depicted in Figure 5.39, and the bit streams V_1 and V_2 can be reconstructed into Nyquist-rate digital samples D_1 by the digital filter in Figure 5.34c. The input sampling switches are bootstrapped [35] to improve linearity, and 4.2 pF sampling capacitors are sized to fulfill the required DR. The quantizer in the first-loop IADC2 is designed as a five-level to lower the two integrators' voltage swing, and DWA circuits [36] are designed to linearize the mismatch error of the multilevel feedback DAC . The coefficient in the second-loop IADC1 with one-bit quantizer

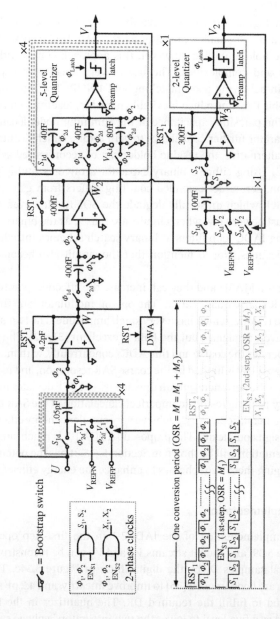

Figure 5.39 The switched-capacitor circuits of the proposed IADC2-1-2. The first-step operation works as an IADC2-1 for 24 clock periods.

is designed as 1/3 to maintain a wide non-overloaded range. In addition to the universal two-phase non-overlapped clocks Φ_1, Φ_2, extra two-phase clocks S_1, S_2 enabled by the control signal EN_{S1} operate the SC circuits for 24 clock periods. The first-step IADC2-1 stores the residue voltage at the last integrator's output node W_3.

To proceed with the fine quantization, the SC circuits reconfigured for the second-step operation are illustrated in Figure 5.40. By disabling S_1, S_2 with EN_{S1}, the last integrator now holds the residue. And, enabling additional two-phase clocks X_1, X_2, with the EN_{S2} turns the whole SC circuit into an IADC2 for another 16 clock periods. Moreover, the kT/C noise contributed by the second-step IADC2 is negligible, and the total sampled thermal noise is dominated by the first-step IADC2-1 when referred to the input of the two steps. In Eq. (5.36), the equivalent quantization error $E_{2\text{-}1\text{-}2}$ can be optimized by selecting an optimal ratio $k_{12} = M_1/M_2 = 24/16$. When M_1 is changed to 12, 24, or 32, the thermal noise penalty factor becomes 1.66, 1.73, and 1.75, respectively, which is not spread widely. The selection of k_{12} is not critical to kT/C noise penalty. The 4.2 pF sampling capacitor used during the first step is also reconfigured and downsized to 0.2 pF, enabling the hold amplifier to drive while reusing the same opamp. The long interconnection from the hold amplifier to the INT1 is only for illustrating circuit operation and should be avoided. In the physical layout of modulators, the three integrators should be placed together. The analog signal nodes and routings should be short, tight, and shielded from noisy digital signals. Switches at the hold amplifier's virtual ground must be carefully designed to avoid leakage current when they are turn-off, especially for very high resolution and low sampling frequency.

Figure 5.41 illustrate the simplified circuits to generate the timing control in Figures 5.39 and 5.40. $EN_{S1}/EN_{S2} = 24/16$ duty cycle and RST_1, RST_2 pulses can be easily generated by a simple counter and logic gates. The non-overlapped clocks S_1 S_2 and X_1 X_2, specifically for each step, are enabled/disabled by EN_{S1} and EN_{S2} with proper driver circuits. These control circuits can easily activate or deactivate the SC signal paths of each step and implement the reconfiguration with minimum circuit overhead. During the M_2 clock periods in the second step, the opamp in INT1 drives less capacitive load and the power consumption can be further saved by using some sophisticated adaptive biasing circuits without increasing circuit complexity.

Conventional multistage $\Delta\Sigma$ ADCs are sensitive to opamp gain requirement and capacitive coefficient matching due to the ECL in the digital domain, as described in Eq. (5.24). In particular, the higher the resolution, the more stringent these requirements become. The SQNR loss versus the matching error of the INT1's coefficient is simulated in Figure 5.42a, and mismatches in each step are plotted, respectively. The matching requirement, while not as harsh as that of the

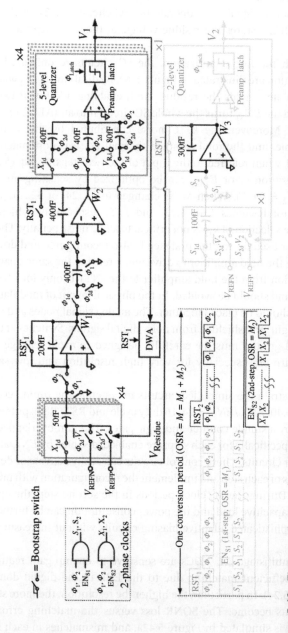

Figure 5.40 The second-step switched-capacitor circuits of the proposed IADC2-1-2. It is reconfigured as an IADC2 operated for 16 clock periods.

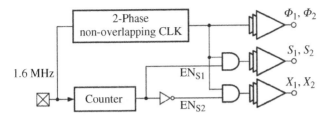

Figure 5.41 Circuits to generate the timing control and two-phase non-overlapped clocks for the two steps.

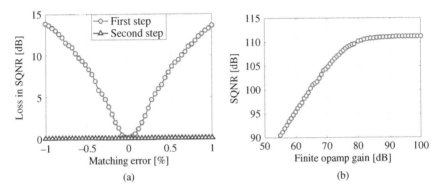

(a)　　　　　　　　　　　(b)

Figure 5.42 (a) Simulated SQNR loss versus matching error. (b) Simulated SQNR versus opamp gain.

conventional MASH $\Delta\Sigma$ ADC counterparts, is more sensitive than single-loop ADCs when very high SQNR is required. The coefficient of 1 is implemented by the ratio of the sampling and integration capacitors, and two 4.2 pF capacitors are used as described in Figure 5.39. It has been confirmed by simulations that 3 dB SQNR loss is caused by 0.2% mismatch. The 0.2% matching requirement is not that challenging, and it can be accomplished by careful layout techniques without additional calibration. For example, common-centroid and symmetric interconnection. The 4.2 pF is reconfigured as 200 fF during the second step in Figure 5.40, and the SQNR loss caused by the matching error is negligible.

Figure 5.42b shows the simulated the SQNR of the IADC2-1-2 versus the opamp gain, where an 80 dB opamp gain can guarantee the 110 dB SQNR. To achieve an 80 dB gain with a wide output range, a two-stage opamp circuit with Miller compensation [37] is implemented and depicted in Figure 5.43. The cascode stage boosts the gain of the first stage and a CT common-mode feedback circuit senses the second stage output and controls the output common-mode level. The opamp's BW designed to cover the dynamic error is the same as that in conventional SC circuits, and more detail in two-step MASH IADC can be found in Ref. [38]. SC

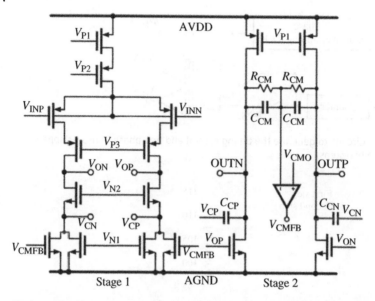

Figure 5.43 The Miller-compensated two-stage opamp in the integrator.

circuits suffer from slewing and consume more power. Using dynamic amplifiers in SC integrators might be more energy efficient than conventional opamps. An example can be found in Ref. [32], which using a floating inverter amplifier (FIA). FIA suffers from lower dc gain and a narrower output range. Applying the FIA in the proposed IADC2-1-2 does not offer benefits. Once the nonidealities are improved in either FIA or the proposed IADC2-1-2, FIA can be a potential solution to reduce the power consumption.

The five-level internal quantizer shown in Figures 5.7 and 5.8 is implemented by using a SC passive adder in front of each preamplifier to sum the feedforward signal paths, and a dynamic latch to digitize the analog signal. The decision is finally stored in an SR latch. The passive adder and comparator circuits are illustrated in Figure 5.44.

As discussed in Section 5.5.3, higher-order IADCs exhibit with more thermal noise. A true IADC5 has a thermal noise 2.8 times that of the fifth $\Delta\Sigma$ ADCs, making the IADC5 less favorable. When implementing a true IADC5 with the same OSR = 40, the sampling capacitor 4.2 pF must be sized as 6.53 pF to maintain the same DR, resulting in higher power dissipation.

5.5.6 Measured Performance

The proposed IADC2-1-2 is prototyped in 0.18 μm technology and powered at 1.8 V supply. The die microphoto of the prototype is presented in Figure 5.45 and it occupies silicon area of 0.48 mm × 0.29 mm, which is only 0.139 mm². With the

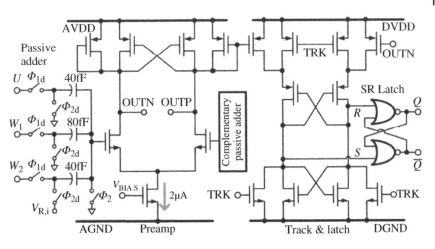

Figure 5.44 The passive adder and comparator circuits.

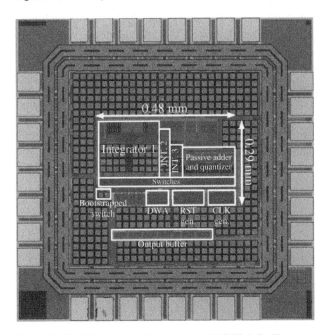

Figure 5.45 Microphoto of the proposed IADC2-1-2 silicon prototype.

total OSR = 40, the IADC2-1-2 is input an oversampling frequency of 1.6 MHz for 20 kHz signal BW. The prototyped is tested with input sine signals at 1 and 17 kHz, respectively, and Figure 5.46 plots the Nyquist-rate measured spectra for each step. When tested with a sine signal at 1 kHz, the IADC2-1 of the first-step (OSR = 24) can measure 66.6 dB SNR, and the noise power is dominated by the

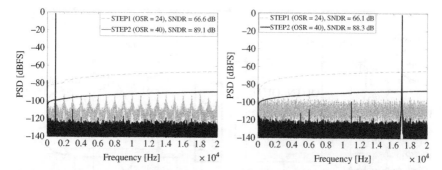

Figure 5.46 Measured spectra for 1 kHz (right) and 17 kHz (left) input sine signal. Spectrum for the first-step IADC2-1 (gray-scaled) and the complete IADC2-1-2 in two steps (black) is plotted, respectively.

third-order shaped quantization error of the IADC2-1. After the fine quantization by the second step, the complete IADC2-1-2 in two steps (OSR $= 24 + 16$) achieves 89.1 dB SNR, which is 22.5 dB higher than the IADC2-1. The noise power of the IADC2-1-2 is dominated by the sampled thermal noise. Measured performance when tested with 17 kHz sine signals is presented in the left of Figure 5.46. The proposed work achieves 89.1 dB maximum SNDR and 107.5 dB SFDR when differential 3 V_{PP} (-1.5 dBFS) is input. Since the five-level quantizer in the first loop must be calibrated, the spectra to compare the feedback DAC DWA turn-on against turn-off is plotted in Figure 5.47. The DWA reduced the harmonic distortions effectively, and it improved the measured SNDR by about 4.3 dB.

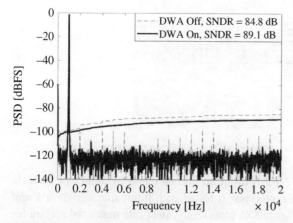

Figure 5.47 Feedback DAC DWA calibration turn-on versus turn-off.

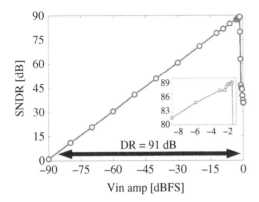

Figure 5.48 Measured SNDR versus input amplitude.

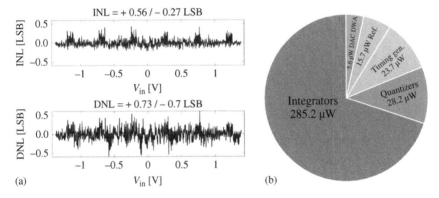

Figure 5.49 (a) Measured INL and DNL. (b) Power breakdown of each building block.

The measured SNDR versus the input amplitude is plotted in Figure 5.48, and the measured DR is 91 dB for 20 kHz signal BW. Figure 5.49a shows the measured INL/DNL errors relative to 14-bit resolution by the histogram testing with a sinusoidal input signal. Supplied at 1.8 V power, the hardware prototype dissipates 358.4 μW. And, the power consumption breakdown for each building block is described in Figure 5.49b. In Table 5.3, the measurement results are summarized and compared with recently published DT- and CT-IADCs. The equivalent Schreier FoMs by DR/SNDR is 168.5/166.6 dB, respectively. Sensor applications to multiplex among several channels usually demand a high-resolution ADC with the shortest latency, very compact area, and excellent energy efficiency. By recycling the same hardware, the proposed IADC2-1-2 demonstrated that it is suitable to interface multiple sensors SoCs.

Table 5.3 Measured performance summary and comparison.

Parameter	[5] This work	[39] ISSCC 2022	[40] TCAS-I 2020/11	[19] JSSC 2022/11	[4] SSC-L 2022/5	[41] ASSCC 2021	[17] SSC-L 2019/9	[42] JSSC 2019/5	[43] JSSC 2019/4	[44] ASSCC 2019	[18] TCAS-I 2015/6
IADC architecture	DT 2-1 2-Step	CT zoom IADC count	DT 3rd interleaved	CT 3-0 DAC cal. free	CT 2-0 IADC + SAR	CT 4th SAR + noise couple	CT 3rd FIR DAC	DT 3rd int. slice	DT 2nd Linear-Exp	DT 2nd Linear-Exp	CT 4th 2-step
Technology (nm)	180	28	180	28	180	180	180	180	65	65	180
Area (mm^2)	0.139	0.014	0.45	0.125	3.99	0.198	0.175	0.363	0.13	0.26	0.337
V_{DD} (V)	1.8	1.2	3	0.9	1.8	1.8	3	3	1.2	1.2	1.2/1.8
F_S (MHz)	1.6	409.6	30	120	64	32	32	30	10.24	128	0.32
Bandwidth (kHz)	20	25	100	1000	1000	200	100	100	20	500	4
Power (μW)	358.4	470	1347	3600	29300	2100	1270	1098	550	20000	34.8
Peak SNDR (dB)	89.1	100.1	85.1	81.2	80.5	78.5	83	86.6	100.8	86.02	75.9
SFDR (dB)	107.5	113.7	101.5	97	86	93	94.3	101.3	–	103.03	88.1
DR (dB)	91	102.2	87.2	89	90.5	82	91.5	91.5	101.8	94.6	85.5
FoM$_{S,\,SNDR}$ [a] (dB)	166.6	176.4	163.8	165.6	155.5	158.3	161.9	166.2	176.4	160	156.5
FoM$_{S,\,DR}$ [b] (dB)	168.5	178.5	165.9	173.4	165.5	161.8	170.4	171.1	175.4	168.5	166.1
FoM$_W$ [c] (pJ/conv.)	0.38	0.11	0.46	0.19	1.69	0.76	0.55	0.31	0.15	1.22	0.85

a) $\text{FoM}_{S,DR} = DR + 10 \cdot \log_{10}(BW/Power)$.

b) $\text{FoM}_{S,SNDR} = SNDR + 10 \cdot \log10(BW/Power)$.

c) $\text{FoM}_W = Power/(2^{(SNDR-1.76)/6.02} \cdot 2 \cdot BW)$.

5.5.7 Conclusion

Implementation of high-resolution IADCs often requires a high-order noise shaping or a high OSR, which degrades the stability and energy efficiency. In this paper, the design of a multistage IADC2-1 is analyzed for a wide input non-overloaded range. To further achieve higher-order noise-shaping performance, IADC2-1-1 is proposed in multistep quantization. The third-order IADC2-1 coarse-quantizes the input analog signal, and then the same hardware is reused to hold the residue voltage and reconfigured into a second-order IADC to continue the fine quantization. The circuit reconfiguration can be simply implemented in the proposed IADC2-1-2. The complete two steps can achieve fifth-order noise-shaping performance with third-order circuits. Moreover, the proposed two-step operation can avoid the penalty of higher sampled thermal noise caused by higher-order IADC weighting.

The hardware prototype is fabricated in 0.18 µm technology which occupies 0.139 mm^2 area. Powered at 1.8 V and a maximum differential input voltage of 3 V$_{pp}$ (−1.5 dBFS), the prototype achieved 91 dB DR and 89.1 peak SNDR over 20 kHz signal BW with 358 µW power consumption. The measured DR/SNDR lead to 168.5/166.6 dB Schreier FoM, respectively.

References

1 V. Quiquempoix, P. Deval, A. Barreto, G. Bellini, J. Markus, J. Silva, and G. C. Temes, "A low-power 22-bit incremental ADC," *IEEE Journal of Solid-State Circuits*, vol. 41, no. 7, pp. 1562–1571, Jul. 2006.

2 Y. Zhang, C. -H. Chen, T. He, and G. C. Temes, "A 16 b multi-step incremental analog-to-digital converter with single-opamp multi-slope extended counting," in *IEEE Journal of Solid-State Circuits*, vol. 52, no. 4, pp. 1066–1076, Apr. 2017.

3 C. -H. Chen, Y. Zhang, T. He, P. Chiang, and G. C. Temes, "A micro-power two-step incremental analog-to-digital converter," *IEEE Journal of Solid-State Circuits*, vol. 50, no. 8, pp. 1796–1808, Aug. 2015.

4 Y. Wang et al., "A hybrid continuous time incremental and SAR two-step ADC with 90.5 dB DR over 1 MHz BW," *Solid-State Circuits Letters*, vol. 5, no. 5, May 2022, pp. 122–125.

5 J.-S. Huang, S.-C. Kuo, Y.-C. Huang, C.-W. Kao, C.-W. Hsu, and C.-H. Chen, "A 91-dB DR 20-kHz BW 5th-order multi-step incremental ADC for sensor interfaces by re-using a MASH 2-1 modulator," *2022 IEEE Asian Solid-State Circuits Conference (A-SSCC)*, pp. 2–4, 2022. doi: 10.1109/A-SSCC56115.2022 .9980590.

6 S. -C. Kuo, J. -S. Huang, Y. -C. Huang, C. -W. Kao, C. -W. Hsu and C. -H. Chen, "A multi-step incremental analog-to-digital converter with a single opamp and two- capacitor SAR extended counting," in *IEEE Transactions on Circuits and Systems I: Regular Papers*, vol. 68, no. 7, pp. 2890–2899, Jul. 2021, doi: 10.1109/TCSI.2021.3077735.

7 S. Ha et al., "85 dB dynamic range 1.2 mW 156 kS/s biopotential recording IC for high-density ECoG flexible active electrode array," in *Proceedings under European Conference on Solid-State Circuits (ESSCIRC)*, Sep. 2013.

8 C. C. Lee and M. P. Flynn, "A 14b 23 MS/s 48 mW resetting $\Delta\Sigma$ ADC," *IEEE Transactions on Circuits and Systems I: Regular Papers*, vol. 58, no. 6, pp. 1167–1177, Jun. 2011.

9 A. Agah, K. Vleugels, P. B. Griffin, M. Ronaghi, J. D. Plummer, and B. A. Wooley, "A high-resolution low-power oversampling ADC with extended-range for bio-sensor arrays," *IEEE Journal of Solid-State Circuits*, vol. 45, no. 6, pp. 1099–1110, Jun. 2010.

10 C. Chen, Z. Tan and M.A.P. Pertijs, "A 1V 14b self-timed zero-crossing-based incremental $\Delta\Sigma$ ADC," in *IEEE ISSCC Digest of Technical Papers*, pp. 274–275, Feb. 2013.

11 G. Mulliken, F. Adil, G. Cauwenberghs, and R. Genov, "Delta-sigma algorithmic analog-to-digital conversion," in *Proceedings – IEEE International Symposium on Circuits and Systems (ISCAS)*, pp. 687–690, 2002.

12 J. Silva, U. Moon, J. Steensgaard and G. C. Temes, "Wideband low-distortion delta–sigma ADC topology", *Electronic Letters*, vol. 37, no. 12, pp. 737–738, 2001.

13 S. Dey, K. Reddy, K. Mayaram, and T. Fiez, "A 50 MHz BW 73.5 dB SNDR two-stage continuous-time $\Delta\Sigma$ modulator with VCO quantizer nonlinearity cancellation", *Proceeding of CICC*, pp. 1–4, 2017.

14 K. Singh et al., "A 14 bit dual channel incremental continuous time delta sigma modulator for multiplexed data acquisition," *VLSID*, 2016.

15 Y. Kwak et al., "A 72.9-dB SNDR 20-MHz BW 2-2 discrete-time sturdy MASH delta-sigma modulator using source-follower-based integrators," *ASSCC*, 2017.

16 L. Shi et al., "A 13b-ENOB noise shaping SAR ADC with a two-capacitor DAC," *MWSCAS*, 2018.

17 M. A. Mokhtar, P. Vogelmann, M. Haas, and M. Ortmanns, "A 94.3-dBSFDR, 91.5-dB DR, and 200-kS/s CT incremental delta–sigma modulator with differentially reset FIR feedback," *IEEE Solid-State Circuits Letters*, vol. 2, no. 9, pp. 87–90, Sep. 2019.

18 S. Tao and A. Rusu, "A power-efficient continuous-time incremental sigma-delta ADC for neural recording systems," *IEEE Transactions on Circuits and Systems I: Regular Papers*, vol. 62, no. 6, pp. 1489–1498, Jun. 2015.

19 M. A. Mokhtar, A. Abdelaal, M. Sporer, J. Becker, J. G. Kauffman, and M. Ortmanns, "A 0.9-V DAC-calibration-free continuous-time incremental delta–sigma modulator achieving 97-dB SFDR at 2 MS/s in 28-nm CMOS," in *IEEE Journal of Solid-State Circuits*, vol. 57, no. 11, pp. 3407–3417, Nov. 2022, doi: 10.1109/JSSC.2022.3160325.

20 I. Fujimori et al., "A 90-dB SNR 2.5-MHz output-rate ADC using cascaded multibit delta-sigma modulation at 8/spl times/ oversampling ratio," in *IEEE Journal of Solid-State Circuits*, vol. 35, no. 12, pp. 1820–1828, Dec. 2000, doi: 10.1109/4.890295.

21 R. G. Lyons, *Understanding Digital Signal Processing*, Third Edition, Prentice-Hall, 2010. ISBN 978-0137027415.

22 E. Hogenauer, "An economical class of digital filters for decimation and interpolation," in *IEEE Transactions on Acoustics, Speech, and Signal Processing*, vol. 29, no. 2, pp. 155–162, Apr. 1981, doi: 10.1109/TASSP.1981.1163535.

23 T. C. Caldwell and D. A. Johns, "Incremental data converters at low oversampling ratios," in *IEEE Transactions on Circuits and Systems I: Regular Papers*, vol. 57, no. 7, pp. 1525–1537, Jul. 2010, doi: 10.1109/TCSI.2009.2034879.

24 J. Steensgaard et al., "Noise–power optimization of incremental data converters," in *IEEE Transactions on Circuits and Systems I: Regular Papers*, vol. 55, no. 5, pp. 1289–1296, Jun. 2008, doi: 10.1109/TCSI.2008.916676.

25 C. -H. Chen, T. He, Y. Zhang and G. C. Temes, "Incremental analog-to-digital converters for high-resolution energy-efficient sensor interfaces," in *IEEE Journal on Emerging and Selected Topics in Circuits and Systems*, vol. 5, no. 4, pp. 612–623, Dec. 2015, doi: 10.1109/JETCAS.2015.2502135.

26 C. Chen, Y. Zhang, T. He, and G. C. Temes, "Two-step multi-stage incremental ADC," *Electronics Letters*, vol. 51, no. 24, pp. 1975–1977, Nov. 2015.

27 J. -S. Huang, Y. -C. Huang, C. -W. Kao, C. -W. Hsu, S. -H. W. Chiang and C. -H. Chen, "A two-step multi-stage noise-shaping incremental analog-to-digital converter (invited paper)," *2020 IEEE 63rd International Midwest Symposium on Circuits and Systems (MWSCAS)*, pp. 158–161, 2020, doi: 10.1109/MWSCAS48704.2020.9184542.

28 A. Agah, K. Vleugels, P. B. Griffin, M. Ronaghi, J. D. Plummer, and B. A. Wooley, "A high-resolution low-power incremental ΣΔ ADC with extended range for biosensor arrays," in *IEEE Journal of Solid-State Circuits*, vol. 45, no. 6, pp. 1099–1110, Jun. 2010, doi: 10.1109/JSSC.2010.2048493.

29 P. Rombouts, W. De Wilde, and L. Weyten, "A 13.5-b 1.2-V micropower extended counting A/D converter," in *IEEE Journal of Solid-State Circuits*, vol. 36, no. 2, pp. 176–183, Feb. 2001, doi: 10.1109/4.902758.

30 J. -H. Kim et al., "A 14b extended counting ADC implemented in a 24 Mpixel APS-C CMOS image sensor," *2012 IEEE International*

Solid-State Circuits Conference, San Francisco, CA, USA, pp. 390–392, 2012, doi: 10.1109/ISSCC.2012.6177060.

31 Y. Chae, K. Souri and K. A. A. Makinwa, "A 6.3 µW 20 bit incremental zoom-ADC with 6 ppm INL and 1 µV offset," in *IEEE Journal of Solid-State Circuits*, vol. 48, no. 12, pp. 3019–3027, Dec. 2013, doi: 10.1109/JSSC.2013 .2278737.

32 Y. Liu et al., "A 4.96 µW 15b self-timed dynamic-amplifier-based incremental zoom ADC," 2022 *IEEE International Solid- State Circuits Conference (ISSCC)*, San Francisco, CA, USA, 2022, pp. 170-172, doi: 10.1109/ ISSCC42614.2022.9731631.

33 B. Gönen, F. Sebastiano, R. Quan, R. van Veldhoven and K. A. A. Makinwa, "A dynamic zoom ADC with 109-dB DR for audio applications," in *IEEE Journal of Solid-State Circuits*, vol. 52, no. 6, pp. 1542–1550, Jun. 2017, doi: 10 .1109/JSSC.2017.2669022.

34 S. Karmakar, B. Gönen, F. Sebastiano, R. van Veldhoven and K. A. A. Makinwa, "A 280 µW dynamic zoom ADC With 120 dB DR and 118 dB SNDR in 1 kHz BW," in *IEEE Journal of Solid-State Circuits*, vol. 53, no. 12, pp. 3497–3507, Dec. 2018, doi: 10.1109/JSSC.2018.2865466.

35 M. Dessouky and A. Kaiser, "Very low-voltage digital-audio ΔΣ modulator with 88-dB dynamic range using local switch bootstrapping," in *IEEE Journal of Solid-State Circuits*, vol. 36, no. 3, pp. 349-355, Mar. 2001, doi: 10.1109/4 .910473.

36 R. T. Baird and T. S. Fiez, "Linearity enhancement of multibit ΔΣ A/D and D/A converters using data weighted averaging," in *IEEE Transactions on Circuits and Systems II: Analog and Digital Signal Processing*, vol. 42, no. 12, pp. 753–762, Dec. 1995, doi: 10.1109/82.476173.

37 P. J. Hurst, S. H. Lewis, J. P. Keane, F. Aram and K. C. Dyer, "Miller compensation using current buffers in fully differential CMOS two-stage operational amplifiers," in *IEEE Transactions on Circuits and Systems I: Regular Papers*, vol. 51, no. 2, pp. 275–285, Feb. 2004, doi: 10.1109/TCSI.2003.820254.

38 M. Akbari, M. Honarparvar, Y. Savaria and M. Sawan, "Power bound analysis of a two-step MASH incremental ADC based on noise-shaping SAR ADCs," in *IEEE Transactions on Circuits and Systems I: Regular Papers*, vol. 68, no. 8, pp. 3133-3146, Aug. 2021, doi: 10.1109/TCSI.2021.3077366.

39 L. Jie, M. Zhan, X. Tang and N. Sun, "A 0.014 mm² 10 kHz-BW zoom-incremental-counting ADC achieving 103 dB SNDR and 100 dB full-scale CMRR," *2022 IEEE International Solid- State Circuits Conference (ISSCC)*, San Francisco, CA, USA, 2022, pp. 1–3, doi: 10.1109/ISSCC42614.2022.9731742.

40 P. Vogelmann, J. Wagner, and M. Ortmanns, "A 14b, twofold time-interleaved incremental ΔΣ ADC using hardware sharing," in *IEEE Transactions on*

Circuits and Systems I: Regular Papers, vol. 67, no. 11, pp. 3681–3692, Nov. 2020, doi: 10.1109/TCSI.2020.3011362.

41 Y. -D. Kim, J. -H. Chung, K. E. Lozada, D. -J. Chang and S. -T. Ryu, "A 4th-order CT I-DSM with digital noise coupling and input pre-conversion method for initialization," *2021 IEEE Asian Solid-State Circuits Conference (A-SSCC)*, pp. 1–3, 2021, doi: 10.1109/A-SSCC53895.2021.9634711.

42 P. Vogelmann, J. Wagner, M. Haas and M. Ortmanns, "A dynamic power reduction technique for incremental $\Delta\Sigma$ modulators," in *IEEE Journal of Solid-State Circuits*, vol. 54, no. 5, pp. 1455–1467, May 2019, doi: 10.1109/JSSC .2019.2892602.

43 B. Wang, S. -W. Sin, U. Seng-Pan, F. Maloberti, and R. P. Martins, "A 550- μW 20-kHz BW 100.8-dB SNDR linear – exponential multi-bit incremental $\Sigma\Delta$ ADC with 256 clock cycles in 65-nm CMOS," in *IEEE Journal of Solid-State Circuits*, vol. 54, no. 4, pp. 1161–1172, Apr. 2019, doi: 10.1109/JSSC.2018 .2888872.

44 B. Wang, S. -W. Sin, U. Seng-Pan, F. Maloberti, and R. P. Martins, "A 1.2V 86dB SNDR 500kHz BW linear-exponential multi-bit incremental ADC using positive feedback in 65 nm CMOS," *2019 IEEE Asian Solid-State Circuits Conference (A-SSCC)*, Macau, Macao, 2019, pp. 117–120, doi: 10.1109/A-SSCC47793 .2019.9056948.

Index

Incremental Data Converters for Sensor Interfaces, First Edition. Chia-Hung Chen and Gabor C. Temes.
© 2024 The Institute of Electrical and Electronics Engineers, Inc. Published 2024 by John Wiley & Sons, Inc.

Printed and bound by CPI Group (UK) Ltd, Croydon, CR0 4YY

16/04/2025

14658573-0003